The Survivalist's Handbook

How to Thrive When Things Fall Apart

RAINER STAHLBERG

Skyhorse Publishing

Parts of pages 10–14 concerning survival without doctors and the advice of Colonel Blanchard C. Henry, MD, are adapted from *Nuclear War Survival Skills* by Cresson H. Kearny (Cave Junction, OR: Oregon Institute of Science & Medicine, 1990).

Part of page 233 discussing the drug trade in small towns is adapted from Justice Richard Neely's dissenting opinion in *State of West Virginia v. Rummer*, 189 W. Va. 369; 432 S.E.2d 39; 1993 W. Va. LEXIS 71 (1993).

Skyhorse Publishing books may be purchased in bulk at special discounts for sales promotion, corporate gifts, fund-raising, or educational purposes. Special editions can also be created to specifications. For details, contact the Special Sales Department, Skyhorse Publishing, 307 West 36th Street, 11th Floor, New York, NY 10018 or info@skyhorsepublishing.com.

Skyhorse® and Skyhorse Publishing® are registered trademarks of Skyhorse Publishing, Inc.®, a Delaware corporation.

Visit our website at www.skyhorsepublishing.com.

10 9 8 7 6 5 4 3

Library of Congress Cataloging-in-Publication Data
Stahlberg, Rainer.
The Armageddon survival handbook : how to prepare yourself for any possible scenario / Rainer Stahlberg.
p. cm.
ISBN 978-1-61608-125-6 (pbk. : alk. paper)
1. Survival skills--Handbooks, manuals, etc. I. Title.
GF86.S72 2010
613.6'9--dc22 2010027069

Cover design by Owen Corrigan
Cover photo credit Thinkstock

ISBN: 978-1-62914-565-5
Ebook ISBN 978-1-63220-143-0

Printed in China

Contents

Photo Credits

Page 312—Oil floats on the surface of the Mississippi River in Louisiana July 24, 2008, following a collision between a fuel barge and a chemical tanker. (U.S. Coast Guard photo by Petty Officer 3rd Class Nick Ameen/ Released; 080724-G-000X-002.)

Page 329—A controlled burn of spilled oil from in the Deepwater Horizon/BP oil spill in the Gulf of Mexico June 9, 2010. (Coast Guard photo by Petty Officer First Class John Masson; 100609-G-5030M-3198.)

Page 331—An oil slick on the surface of the Gulf of Mexico, May 7, 2010. (U.S. Air Force photo by Tech. Sgt. Adrian Cadiz/Released; 100509-F-0848C-043.)

Introduction

I n 1651, Thomas Hobbes published the *Leviathan*. In this book, he introduced the concept of the "state of nature," a description of a state of pure anarchy. Hobbes says life of man is "solitary, poor, nasty, brutish, and short." That is, each man must provide for himself all things, including defense. In this state of pure anarchy, one cannot even sleep for fear of never waking up.

The solution was to introduce some other kind of state, be it tribe, nation, empire, or dictatorship. Today there are some states that are in anarchy, but they are the exception. States were invented because they are very difficult to kill. Unlike individuals, states never sleep. Yet within each state, the individual is important, particularly if that individual is you. This book was written to help the individual rise to a statelike security.

The coming years will divide America by lifestyle . . . income . . . outlook . . . and even geography. Which side will you end up on? The haves and the have-nots will grow light-years apart, separated by a technological wedge that is about to be driven right through the heart of the middle class, with the world drifting apart on religious, ethnic, and age lines. Just look at Ireland, Israel, and France. Your first duty is to *stay alive*. You are of no use dead.

You know, it does not take any special genius to end up on the right side. All it takes is a determination to look beyond the headlines and the willingness to seize opportunities. Suppose, for example, that a couple decades ago a person had come to you with a blueprint of the years ahead—specific, matter-of-fact details about major events to come. It explained how the U.S.S.R. would fall apart and that they would dismantle the Berlin Wall. Yugoslavia and parts of the former U.S.S.R. would erupt into civil war. It would explain how the dot-com boom would bring unprecedented financial gains before a catastrophic crash and how just a few years later the crash of the U.S. housing market would send countries around the world into economic upheaval. Real estate investors (and the rest of the world) would face a disaster. It would explain how a Texas governor and son of a former president would become president of the United States, in the two most controversial elections in American history. It would tell of the attacks on September 11, 2001, the collapse of the twin towers, and the two wars that the United States would become entrenched in as a result. Finally it would explain that the nation's first African American president would be sworn in on a cold morning in January 2009.

Chances are you would have looked at these predictions with much skepticism. Because when you look back, the world was a much different place. Just think how far ahead you would be right now if you had that glimpse into the future. For one thing, you would own a self-sufficient retreat paid for by your shares in Google.

This is why we urge you to pay close attention to what is happening in the world today. Some scenarios detailed in this book may seem farfetched, but remember, the forecasts of the last ten years seemed really farfetched back in the 1990s.

One word of warning: Some of what you are about to read will not please you. You may not want to believe it. Once-beautiful cities will be destroyed. Formerly rich suburbs will turn into crime-ridden slums, with a lower quality of life than Third World nations and sinking life expectancies.

But remember that this is only half the picture. There is another future—a future in which people reading this book live longer and in style, too. A future where safety and security is once again provided by your immediate community. You must begin to use self-help in crime control because government both cannot and will not protect us under any circumstances conceivable in the real political world.

Most Americans are totally unprepared to make the shift to the future. There is a lack of education, training, and gumption. Seduced by the promises of big government, big unions, and big business, they think the world owes them a living. When disaster comes and

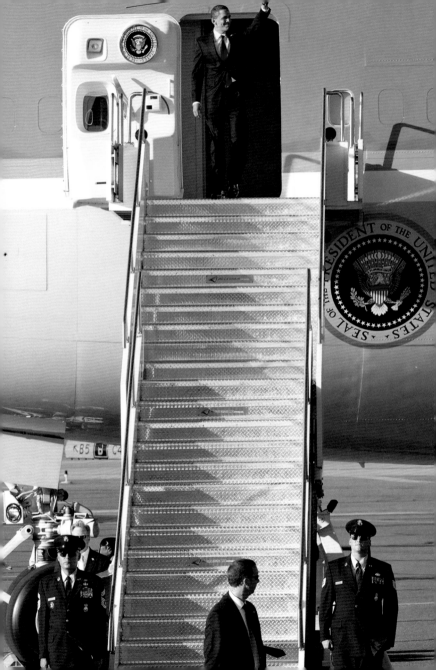

government, unions, and business can't afford to help them muddle through, many people will not be able to cope. Even worse, they will simply die. Many people would much rather die than think. You will see proof of this statement in almost all scenarios that will be presented.

You must accept the fact that you cannot prepare for everything. Only the government can afford to do that. Prepare only for those scenarios likely to happen in the next three to five years in your specific area.

For instance, where *I* live, I can expect:

* Severe winter weather
* Political instability due to breakup of a country
* Heavy thunderstorms
* Chemical spills
* Conflict from rat packs

Our society is facing emergencies on a daily basis. They are emergencies because we are not prepared for them. These range from economic, climatic, ecological, military, terrorist, all the way to a major depletion of the ozone layer. There are books upon books on how to cope with this disaster or the other and how to mitigate their effects. Then there are works of fiction that deal with different scenarios. But there has been a need for a concise, one-volume *checklist* to use as a logical, sequential guide on what to do, when. And that is precisely what *The Survivalist's Handbook* is.

In the course of preparing the scenarios, I used Herman Kahn's advice stating that *we must think about the unthinkable*. Unfortunately, if you follow through any scenario to its conclusion, you will feel quite depressed. This book is not for the fainthearted, but I had to carry the scenarios through to their conclusions in order to be of help to you. With a little thinking about them, you will be able to form a scenario applicable to your circumstances or your area. Preparedness is the name of the game. Most people are not even prepared for a good storm.

What you will not find is the gentle art of writing memorable suicide notes because you had a bad-hair day, car troubles, divorce, or even bankruptcy. Forget about packing in the crowds for your wake. What is covered here is how to stay alive and thereby ensure that your family tree is not pruned in a big way. Survival is a very low-key business. All you have to do is stay alive and keep quiet.

The keys to survival are knowledge and skills. The time to get these is now, not when the balloon goes up. Books can help you with the knowledge. However, skills can only come from practice. Before you practice, you will have to know what to practice. This book will help in that department, too.

Get a copy of *The Encyclopedia of Survival*, an unusual book in that it is aimed for long-term survival. It also presents topics in an unusual way. If a subject is covered extensively in several books in the local library, the subject is summarized. But if information is not readily available, it is given in step-by-step detail.

Finally, the psychology of true survival revolves around one sentence: *Never give up*. What does it mean? Your mind must provide alternatives. If your assault rifle gives up, you can always make a single-shot muzzle-loading pipe gun. And when that is gone, there is always the possibility of making a crossbow. Then finally, there are your hands.

The advice "never give up" goes further than just equipment. It should become a lifestyle for you and your family. You should look at any piece of news, government regulation, change in your neighborhood, or technological advance with the view of how to convert it to your advantage. This is a continuing process. You cannot make your plans and preparations and then relax, for the situation changes constantly.

Survival of what? Whatever happens is the answer. Webster's dictionary defines "survival" as "the continuation of life or existence in the presence of or despite unusually difficult conditions." Note that we are not talking about conquering, winning, or even overcoming. We are talking about simply keeping alive. Under some circumstances, this can be a tall order.

What are the essentials? First comes the attitude to survive. Then comes breathable air. Without air, you would last only a minute or two. The third essential is potable water. Without it you would not last much beyond three days. For long-term storage, you should have a supply of sterile or disinfected water. To disinfect water, to each gallon, add sixteen drops of liquid chlorine bleach (Clorox or Purex-type bleaches, containing four to six percent sodium hypochlorite). One teaspoon of bleach disinfects ten gallons of water. Then there is food, shelter, protection, clothing, and a host of other vital items. One purpose of this book is to show you what you will need to do without, what you can make, use as substitutes, or scavenge.

Almost all scenarios recommend that you have food on hand—in some cases, enough for a week or two; in others a year's supply is suggested. Start to build your food stockpile now! Add to it as you can afford it. Having food will enable you to sit out the immediate panic associated with emergencies.

Many otherwise clearheaded people can't accept the decline in our social systems

General Survival

as a result of climatic change, overpopulation, reduced resources, government deficits, and other causes. This has happened before. Just read history books. Civilizations have come and gone. We have had Ur in Chaldea, Babylon, the Medes, the Roman Empire, and even perhaps the famed Atlantis. Yet some people have survived all of these declines. (Even Atlantis. There have been too many mysterious finds that show this large island was more than a myth and that there were those who reached other civilizations.)

In most cases, survivors were rural dwellers. City people are the first affected, and in our society, we are looking at 80 percent of the population being urban dwellers. Do not rely on your gold supply to get you food in the initial period of a panic. Remember you can't eat gold!

There are four steps to making a disaster plan. These are:

1. Make a list of disasters than can happen to you.

This is where this book can be of the most use to you. As part of your preparations you should:

- List the types of disasters that can happen in your area.
- Take a first-aid course and learn about animal care after a disaster.
- Find out about local warning signals and plans.

2. Create a disaster plan.

This plan is partially prepared for you on the following pages. You should sit down with your family and friends and fit the ideas provided in this book into a *site-specific* plan.

- Discuss what to do in different situations.
- Prepare an evacuation plan.
- Decide on places and contacts to reunite your family and your group.

3. Put your plan into action.

- Have emergency numbers with each member of your group.
- Review your insurance coverage.

- Accumulate equipment and supplies.
- Install smoke detectors and have ABC-type fire extinguishers in your home.
- Find the escape routes from your home and safe spots in the house for different types of disasters. List the hazards in your home as part of this preparation.

4. Update your plan and conduct drills.

- Update plans as the situation changes or every six months.
- Test your equipment frequently, and rotate batteries and perishable supplies.
- Replace stored water every three months.

Throughout this book, references are made to the government as an alien entity. In theory, *we are the government, you and I*. Alas, that is not so in practice. What happens to honest people of high ideals when they are elected into office is an age-old mystery. The net result of the difference between what the government says and what it does has created a deep mistrust in all forms of governments by intelligent people. *The Armageddon Survival Handbook* is written for intelligent people.

We do not place too much reliance on the law enforcement authorities. Why is that? Although the United States made short work of the Iraqi army, it has been less successful with criminal elements at home. Why? The answer to that question is the result of complex social, political, and economic forces at work. The bottom line, however, is that lack of money is not the only factor that makes law enforcement agencies incompetent.

Our law enforcement is also incompetent because all of us, at some point in the system, whether consciously, intelligently, or not, want it to remain incompetent. That may be a counterintuitive proposition, but it is, nevertheless, true. Why do we have an FBI, DEA, BATF, border patrol, customs service, and secret service? Not to mention the park police. According to some early historians,

the ones who joined the police were too scared to steal and too lazy to work. In the nineteenth century, when Robert Peel set up the English police, Parliament maintained that a good police system was a threat to civil liberties! 'Nuff said.

The scenarios laid out were chosen from those envisioned by intelligent people asking "What if?" questions and then sitting down to write what they would do to survive that particular scenario. Some of these scenarios seem farfetched. But remember, when we look back, the future always seems farfetched. This, coupled with our normal reaction of wait and see, can be detrimental to our well-being under most circumstances.

Coping or surviving requires an open mind and a willingness to change. The human species is very adaptable. This is why we are in a dominant position today, and this is in spite of our lack of fur, large teeth, or speed.

Unfortunately, most of us grow up in urban areas. As a result, we have no idea how to grow food, skin a game animal, or even how to repair a tractor. Should we be thrust back into the Middle Ages, most of us would be utterly helpless and perish. The early twentieth-century armies preferred farm boys, since a farmer in those days had all the skills of a carpenter, builder, mechanic, plumber, and machinist rolled into one. In contrast, today we have much greater specialization. This specialization is required to keep our civilization functioning. However, in a survival scenario when we are thrust back to primitive conditions, this specialization works against us. Most of us do not even know how to start a fire without matches.

To be a survivor, you should at least have the skills of a scout, take a first-aid course, know something about researching a topic, and have access to books dealing with the technologies of the 1850s to 1940s. You can amass a library by going to garage sales and used-books stores. Do not forget to pick up back copies of magazines like *Harrowsmith, Mother Earth News, Popular Mechanics, Popular Science, Backwoods Home Magazine,* and *American Survival Guide.* It is also a good idea to have as many catalogs as you can collect.

Catalogs show all kinds of accessories for equipment and serve a useful function in showing you what can be done with items picked up at garage sales.

A related issue is other people cooperating with you. We are *social* animals, that is to say that a group of humans will possess more knowledge and skills than an individual. At least that is the way it should be. So get to know the people around you, but do not preach to them. You don't want to be labeled a kook.

Another important aspect is adaptability. Our species is very adaptable, according to historians. Looking around today, this may not seem to be so. We have a plethora of specialists in every profession, some so specialized that it boggles the mind. As a survivor, you cannot focus only on food, guns, or whatever. You must plan to know about and have the supplies to provide potable water, food, shelter, protection, tools, first-aid skills and equipment, fire-making supplies, and transportation to fulfill your needs. The potential survivor must be nonspecialized and all-encompassing at the same time. Be ready to change occupations to take advantage of the situation.

Another important point: Deal in cash only! Do not leave a paper trail behind of what you buy or what you sell. Use a credit card when renting a car or checking into a hotel. But when you return the car or check out of the hotel, pay with cash. Remember, banks microfilm all records. Someone rooting around can find out your past purchasing patterns. Once found out, you may have a lot of explaining to do. For example, why in the world would you want a gillnet when they are not legal for fishing?

Another word about cash. Do not make yourself conspicuous by using large, denomination bills for small purchases. You don't want the store clerk to be able to describe you in detail later.

Remember that accumulating supplies is easy only before Day One. After that, it is a chancy and more expensive proposition. Many people will be after the same materials once the situation becomes sticky. So stock up early. And keep quiet about what you buy. Cache some of your supplies using lengths of plastic sewer and water pipes. The ends can be easily sealed with end caps available from the same source as the pipes. Governments have a nasty habit of confiscating supplies from people who prepared for disasters to give to the indigent. Their logic runs to the philosophy that by giving supplies away, you reduce the possibility of the indigents turning into predators. All they are doing is putting everyone at risk.

Once you have to forage for supplies, a city directory or even the yellow pages are of great assistance to you. Having these on hand, you can organize scavenging tours for the supplies you need. Another source of information for supplies may be found in trade directories. These are available at most libraries and the purchasing departments of corporations. You should be aware of what is available in your community and the surrounding area. When you forage for supplies, do not overlook company lunchrooms and truck stops. Other sources can be abandoned trucks and vehicles, abandoned farms, and even museums.

It is important to have a designated place for your group to meet should you have to evacuate. You must have a place picked out well before the excrement hits the fan. It should be easy to identify and to reach by all members of your group. And it should not be on the official evacuation route.

A few words about surviving without doctors. Most physicians, hospitals, and medical supplies are located in cities. Most disaster scenarios will hit cities harder than the countryside. Therefore, you may have to do without conventional medical treatment after a disaster. The human body has remarkable capabilities of healing itself, especially if the injured and their companions practice intelligent benign neglect.

Colonel Blanchard C. Henry, MD, an authority on mass evacuations and treatment, recommends the following rules for survivors under primitive conditions:

- Wounds: Apply only pressure dressings to stop bleeding.
- Infected wounds: Do not change dressings frequently. The formation of white pus shows that white corpuscles are mobilizing to combat the infection. In World War I, wounded soldiers

in hospitals suffered agonies having their wounds cleaned and dressed frequently. Many died as a result of such harmful care. In contrast, before antibiotics became available late in World War II, casts and dressings on infected wounds sometimes were not changed for weeks.

- Pieces of glass deeply embedded in flesh: Do not probe with tweezers or a knife in an attempt to extract them. Most glass will come out when the wounds discharge pus.

- Burns: Do not apply grease, oil, or any other medication to the burned area. Cover the area securely with a clean, dry dressing or folded cloth. Do not change the dressing frequently. For most burns, the bandage need not be removed until the tenth to fourteenth day. Give plenty of slightly salted water: about one teaspoon (4.5 g) of salt per quart (or liter), preferably chilled, in amounts of one to three liters daily.

- Broken bones: Apply simple splints to keep the bones from moving. Do not worry about deformities. Most can be corrected later by a doctor. Do not attempt traction setting of broken bones.

- Shock: Keep the victim warm. Place blankets or other insulating material under him. Do not cover him with so many blankets that he sweats and suffers harmful fluid losses. Give him plenty of slightly salted water.

- Heat prostration: Give adequate fluids, including slightly salted water.

- Simple childbirth: Keep hands off. Wait until the mother has given birth. Do not tie and cut the cord unless a potent disinfectant is available. Instead, use the primitive practice of wrapping the cord and the placenta around the infant until they dry. Avoid the risk of infecting the mother by removing the rest of the afterbirth. Urge the mother to work to expel it.

- Toothache: Do not attempt to pull an aching tooth. Decaying teeth will abscess and fall out. This is a painful but seldom fatal process—one that was endured by most of our remote ancestors who reached maturity.

While on the subject of medical matters, do not forget about veterinary antibiotics. People who for decades have used antibiotics to combat their infections have not produced normal quantities of antibodies and have subnormal resistance to many infections. People who have not been dependent on antibiotics have these antibodies. In the aftermath of a disaster, most survivors would be in rural areas. Many would need antibiotics. Much of their need could be met by supplies of veterinary antibiotics kept on livestock and chicken farms, at feed mills, and in small towns.

Many animals are given more antibiotics in their short lives than most Americans receive in theirs. In many farming areas, veterinary antibiotics and other medicines are in larger supply than are those for people. Realistic preparations to survive should include using these supplies.

If you can, have a retreat. The retreat should be off the main roads. Be prepared to chop a few trees to control access to it when the time comes. Keep it low profile and located close to a source of water. Your retreat should have arable land and a source of firewood nearby. Stock it with food, tools, firearms, and ammunition for long-term self-sufficiency. If you do not live in your retreat, cache your supplies away from the building. A buried cache is hard to beat.

A few words about surfing the Web. Anything posted on the Internet leaves a permanent trail. If you were a foolish youth and sent nasty messages concerning issues you have disagreements with, it can come back to haunt you five years later when you run for city council. Think about it.

Another point: While you are on the Net, some very sophisticated programs can probe the contents of your computer. General advice: Never put anything on the Net you wouldn't want to read in tomorrow's newspaper. False names and other privacy measures work only up to a point.

The scenarios laid out on the following pages are the ones most likely to happen in the near future—yes, including the visit by aliens. The one constant through the ages is that change is ever present.

Who could have imagined YouTube ten years ago, the iPhone, 9/11, or the rising distrust in the federal government's desire to safeguard the Constitution?

How to Recognize the Arrival of Day One

To some, Day One means the mushroom cloud on the horizon, the choking in the Tokyo subway, or trekking across subarctic areas in running shoes while the temperature plummets to −40°F. Wrong, that is *Day Two*! By that time, you should have been in a fallout shelter, avoiding the subway, or wearing insulated clothing and pulling a cargo toboggan.

To recognize Day One, you must be an armchair general, keeping abreast of the news and having the knowledge and equipment to muddle through Day Two. The only way to survive in style is to have a plan and the knowledge to carry it through. Day One can be anywhere from one day to ten years long, depending on the disaster scenario you are facing. In the individual scenarios, we will try to focus on their estimated duration.

Similarly, Day Two or Day Three are not necessarily of a twenty-four-hour duration. Perhaps it would have been more precise to call them phases. However, it is easier to think of them as days. When you look back upon an interesting phase of your life, you think of specific days while skipping over other intervening ones when nothing much happened. So it is with this book.

Day One is critical to your well-being. This is when you plan, prepare, accumulate supplies, and learn new skills to carry you through the following days. What you do on Day One determines whether you will live through any of the scenarios detailed in this book in style. Some people will survive without preparation by sheer luck, but that is not the way to bet. What you are betting is your existence and the existence of those you love.

Day One is unusual in that if you do not prepare for it, unless you are very lucky, you have lost the battle. Your planning starts when you realize that things can happen to you. Day One is the day when many people will ridicule you if they know that you are preparing for something. On the other hand, if you do not prepare, you are not likely to have any descendants. As the old saying goes, "If your parents did not have any children, you are not likely to have any either."

At the very least have your passport and other important papers up to date and close to you. What should you keep in your safety deposit box? Your fire insurance policy, copies of important papers, a copy of your last will, and similar documents. Do not keep cash, gold, firearms, or like items in your safety deposit box. Under many circumstances, they will get you into trouble. Even in a nonemergency situation, think what the IRS, the BATF, or the FBI would make of them.

Where should you keep the originals of your papers and your valuables? The answer depends on where you live. If your home is one room in a boardinghouse and your landlord regularly pokes around your place, the best solution is to carry your valuables with you, wherever you go. On the other hand, if you own a house in the suburbs, there are many more places to hide your valuables and supplies.

There is a tendency among us to brag a little about the preparations we make. I can't repeat it often enough—keeping a low profile is very important in this business. You may have an impromptu show-and-tell session with a neighbor, who doesn't believe there could be hard times. Once something does happen, he will turn on you for your supplies. His lack of preparedness is partially due to the government-encouraged belief that "it can't happen here." It will and you better be prepared for it, even if your neighbor isn't.

Except in a few cases, Day One will not arrive with large neon signs and sound effects proclaiming that it is here. In most cases, Day One arrives like a thief in the night. Sometimes it is only on

Day Two that many will realize that Day One was yesterday. As a general piece of advice, assume that Day One is on hand now. Lay in supplies, learn new skills, and keep informed on what is happening in the world. Accumulate stocks of food that you eat now, and rotate those stocks. Do not rely too much on your freezer and refrigerator. Electricity is one of the first things to go in an emergency. Rely on canned foods, dehydrated sachets of soup and pasta dishes, canned meats, stews, and other nonperishables.

Day One has already come for some scenarios, and it is very close to many others. We are living at the edge, and given the current state of affairs, we should be on guard. Once a scenario unfolds, it can progress at a frightening pace. Be prepared at all times.

What to do:

- Take stock of what you are eating, and take stock of what you have on hand. That will tell you how long you could survive if the stores close tomorrow.

- Where do you get your water? Find what other sources you have, and have those sources tested.

- What happens to your sewage? How would you cope if your system is disrupted?

- Are you on medications? How long could you last without them? Do you have an alternate source for them? Always refill your prescriptions ahead of time. If questioned, just say that you are taking a trip. Have at least a month's supply on hand.

- Is your vehicle ready to roll in case of an emergency? How much gas do you have? What about lubricants, brake fluid, spare parts?

- Do you have a place to go? Do you know the topography of the area between your home and your place of retreat?

- Do you have the knowledge to deal with emergencies? Do you have the skills to put the knowledge into practice?

- How can you make your home harder to break into by criminals and looters?

- Can you handle firearms for self-defense? Do you have the fire-arms along with ammunition and spare parts to maintain them? Can you reload ammunition?
- Have you taken a first-aid course in the last five years? Do you have the supplies and instruments to give first aid?
- Are you aware of what is happening in your community, your country, and in the world? Do you have radios, communication devices, and newsletters to keep you informed?
- Do you have the financial or barter assets to start all over again?

A lot of hard questions are raised above. Yet, unless you sit down and answer them in the hard light of honesty, your chances of surviving even a temporary interruption in your present mode of life will be reduced.

Then there is the question of security while you are preparing to survive. The ten commandments of security are:

1. Do not discuss personal or family business with anyone not directly involved.
2. Do not trust a politician or bureaucrat's word or promise.
3. Never give your real name or address when purchasing survival supplies.
4. Never let strangers into your home.
5. Do not turn your back on an unlocked door or window.
6. Do not have your street address on any of your IDs or mail. Use a P.O. box as much as you can.
7. Do not keep all your money or valuables in the same bank. Have several bank accounts under different names.
8. Never rely on someone else doing anything correctly or on time.
9. Do not have important mail sent to your home.
10. Always set the alarm and lock your garage when leaving your home.

You should be getting concerned by now. I would suggest that while you still have time, learn and practice new skills and lay in supplies and books. You must have the right psychology to survive.

How to Cope Day by Day

Now that you can recognize Day One, let us look in a general way at coping with the following days.

As a rule of thumb, while the federal and state governments continue to function, they will attempt to deal with a situation in the traditional ways. To know what these measures will be, you must study history. Most likely, the government's first step will be to declare a state of emergency with its accompanying declaration of martial law. As the situation deteriorates, you are likely to see restrictions on movement, ration books for food, fuel, and other items, and a host of other restrictive regulations. Some portion of the population may be placed in custody as a precautionary measure. These people will be deemed to be opponents of the government.

This is just the tip of the iceberg. The other nine-tenths of it will be the actions of the affected population. Will they perceive it as a short-term event and cope with it, or will they react in sheer panic? As a rule, you should not expose yourself and your family to the panic reaction of either the population or the government. This requires information, communication, planning, and supplies to sit out the initial panic period.

At the same time, there are several scenarios that call upon you to evacuate immediately. This is commonly called "bugging out." To decide whether to stay put or bug out, study the scenarios.

Waiting out in a low-profile manner enables you to see the events unfolding and watch the crazies and the unprepared do each other in. This period is not without its inherent dangers. For example, if your neighbors know that you stockpiled food and other necessities, they may want to raid your supplies. That is why you must do every preparatory action in a low-key way. Remember, the best camouflage is to blend in with the background. If your background is a suburban area, blue jeans and the seemingly beat-up four-wheel drive you use for fishing is more appropriate than wearing tiger stripes and driving a war wagon.

The typical human reaction is to pigeonhole people by their appearance. Once categorized, they are thought of as groups. You certainly don't want to be pigeonholed as a prime target for the sniper in the police SWAT team. Therefore, camouflage yourself so that the eye looking for a target skips over you and starts to look for a "real target." Once again, blend in with your background, both in appearance and mannerism. To some this may sound like conformity. It is not. What I propose is *seeming* conformity.

As you read on, you will find repeated references to a small "bug-out" kit. Having a kit of this nature is your first step to survival. Such a kit should be personalized to your needs. A base-camp sample is described in the next section and a small one in Chapter Four. Have a kit for each member of your family. If you have children, keep in mind that they can carry less, and you must replace certain items of clothing and footwear as they grow.

Then there is the idea of forming a cooperative—which is different from a militia. This should be done on Day Two or Three. By that time, the situation has deteriorated to the point that a group of like-minded people is needed. Don't form any alliances earlier than this, except in an informal way. The reasons for this are manifold.

For one thing, if nothing happens for years and you have a communal group or cooperative, it is likely to arouse the curiosity of federal, state, and local police forces. You really do not need that kind of attention. Furthermore, people move, change their lifestyles, or divorce. You don't want ten percent of your shelter space to become the star attraction in a messy divorce settlement.

Moreover, groups formed for benign purposes sometimes do really stupid things because they are a group. For example, look

at some of the so-called militias trying to take on the forces of the federal government. No civilian group can stand up to the government's might in a head-on confrontation. As said earlier, you don't want to conquer, you just want to survive as best as you can.

Day One

What to do:

- You must have the initial supplies to enable you to carry on through Day Two and beyond. A bug-out kit is a vital piece of equipment. Having one will enable you to react on a moment's notice to changing conditions.

- You should keep informed. Today there is no excuse for being ignorant. The information you can get from a shortwave radio, television, and magazines—not to mention the Internet—will enable you to keep up with what is happening in the world and how those events might affect you. Civil strife in Sri Lanka may impact on your tea supply, to say the least.

- There are certain supplies that are very cheap today. However, should there be an interruption of transportation networks, they will be very highly priced. Salt and pepper are two of these commodities.

- Find like-minded people and work together to have a support group around you when disasters happen. Try to persuade them to have supplies on hand.

- Have enough fuel on hand to get you to your chosen retreat or predetermined assembly point. Store it away from your house and remember to rotate the fuel. Gasoline should be rotated every six months and diesel every nine to twelve months.

Day Two

To cope with the situation day by day depends on the scenario. Go to the appropriate scenario. As the situation evolves, you should be able to recognize what is happening. You will find that the scenarios

12 oz Fill Line

FIRST STRIKE RATION

FIRST STRIKE Rations are
Formulated to Help You
Perform at Your Peak and
Complete Your Mission!

TEAR POUCH AT NOTCHES. OPEN ZIP
ADD 6 OZ WATER (1/4 CANTEEN CUP
TO FILL LINE. CLOSE ZIPPER,
SHAKE TO MIX.
SINGLE USE ONLY.
CONSUME PROMPTLY.

laid out will have to be changed as the situation unfolds. Do so. The contents of this book are not carved in stone.

Survival Equipment Checklist—Base

Food and cooking equipment:

☐ For a one-year dietary supply you will need, at a minimum:

Wheat	300 pounds
Powdered milk (nonfat)	100 pounds
Sugar	50 pounds
Salt	5 pounds
Multiple vitamins	365 tablets

☐ Food, canned and sealed containers. Precooked. 2-month supply

☐ Food, dehydrated and freeze dried. 4-month supply

☐ Food, bulk, cereals, vegetables. 6-month supply

☐ Salt, ½ tablespoon per person per day

☐ Multivitamin supplements

☐ Seasonings

☐ Sugar

☐ Tomato powder

☐ Onion flakes

☐ Bouillon cubes

☐ Yeast

☐ Baking powder

☐ Baking soda

☐ Milk powder (Test before laying in a supply. Some brands may not agree with you.)

☐ Butter, shortening, margarine, and fats

- ☐ Cheese powder
- ☐ Knives, forks, and spoons
- ☐ Cooking utensils
- ☐ Paper plates and cups
- ☐ Paper napkins
- ☐ Pots and pans
- ☐ Emergency stove or camp stove
- ☐ Fuel for stove
- ☐ Matches or lighters
- ☐ Nonelectric can and bottle openers
- ☐ Coffee and tea
- ☐ Cooking thermometer
- ☐ Special dietary supplements
- ☐ Aluminum foil
- ☐ Plastic bags
- ☐ Plastic-bag sealer
- ☐ Soap for dishes
- ☐ Wheat grinder
- ☐ Spare grinding stone
- ☐ Flour sifter
- ☐ Measuring cups
- ☐ Scrubbing brush
- ☐ Seeds, nonhybrid
- ☐ Gardening equipment
- ☐ A double boiler (This can be used for making chemicals, too.)
- ☐ Pet food, if you have pets

Water supplies:

- ☐ Water, 4 quarts per person per day for drinking and cooking
- ☐ Water filter, best to have one with both ceramic filter and activated carbon element
- ☐ Spare ceramic filter and activated carbon for water filter
- ☐ Household bleach, iodine tablets, or other purification tablets
- ☐ 5-gallon plastic containers

☐ Water canteens (In the north, have stainless steel canteens.)

Sanitation equipment:

☐ Plastic buckets
☐ Plastic garbage bags
☐ Bleach or other disinfectant
☐ Toilet tissue
☐ Mirror
☐ Razor
☐ Shaving cream and aftershave lotions
☐ Toothbrush
☐ Toothpaste
☐ Soap
☐ Detergent
☐ Laundry soap
☐ Shampoo
☐ Comb
☐ Scissors
☐ Paper towels
☐ Dry wipes
☐ Sanitary napkins/tampons
☐ Towels
☐ Face towels

Shelter, camping equipment:

☐ Sleeping bags
☐ Sleeping-bag liner, overbag, and sleeping pad
☐ Cots
☐ Plastic tarps or poncho
☐ Flashlight
☐ Spare batteries and bulbs
☐ Battery-operated radio, shortwave preferred
☐ CB and scanner radios
☐ Spare batteries for the radio

- ☐ Candles
- ☐ Clock or watch and calendar
- ☐ Pen, pencil, paper, notebook
- ☐ Cards and games
- ☐ Binoculars
- ☐ Indoor/outdoor thermometer
- ☐ Fire extinguishers
- ☐ Kerosene heater
- ☐ Kerosene
- ☐ Lantern
- ☐ Composting or chemical toilet
- ☐ Tent

Books:

- ☐ Survival manuals such as *Encyclopedia of Survival*
- ☐ Herb identifier handbook
- ☐ Manuals for firearms and other equipment
- ☐ Medical books such as *Where There Is No Doctor*
- ☐ Do-it-yourself books on back-to-basics subjects
- ☐ Older technical books dealing with obsolete technology
- ☐ Road atlases and maps
- ☐ Bibles, religious materials
- ☐ Cookbooks, particularly those dealing with dehydrated foods

Tools:

- ☐ Shutoff wrench for gas and water
- ☐ Saw
- ☐ Ax
- ☐ Screwdrivers
- ☐ Pliers and wrench
- ☐ Auger
- ☐ Hammer
- ☐ Crowbar, wrecking tool
- ☐ Shovel or entrenching tool

- ☐ Pickax
- ☐ Bolt cutters
- ☐ Machete
- ☐ Files, rasps
- ☐ Oils and lubricants
- ☐ Plastic sheeting
- ☐ Masking tape
- ☐ Duct tape
- ☐ Dust masks
- ☐ Glues
- ☐ Buckets
- ☐ Rope
- ☐ Rubber hose
- ☐ Wire
- ☐ Aluminum foil
- ☐ Nails and screws

Personal equipment:

- ☐ Backpack
- ☐ Sewing supplies, needles, thread, spare buttons
- ☐ Safety pins
- ☐ Clothing
- ☐ Change of underclothing and socks
- ☐ Boots
- ☐ Foot powder
- ☐ Hat
- ☐ Spare glasses, contact lenses, hearing-aid batteries
- ☐ Special medications (insulin, etc.)
- ☐ Sunscreen
- ☐ Sunglasses
- ☐ Maps and compass
- ☐ Fishing line and hooks
- ☐ Snaring wire
- ☐ Space blanket

- ☐ Whistle
- ☐ Pocketknife—best to have two, one a Swiss Army type and the other a good quality folding or belt knife
- ☐ Magnesium match or equivalent
- ☐ Matches
- ☐ Magnifying lens
- ☐ Binoculars
- ☐ Whistle
- ☐ Strobe light
- ☐ Signaling mirror
- ☐ Flashlight
- ☐ Spare batteries and bulbs
- ☐ Watch with altimeter
- ☐ Water canteen
- ☐ Water-purification tablets
- ☐ Candy, chocolate bars
- ☐ Trail rations
- ☐ Fishing kit

Defense equipment:

- ☐ High-power rifle
- ☐ Ammunition for rifle, 1,000 rounds
- ☐ Sling for the rifle
- ☐ Ammunition pouches
- ☐ Scope for the high-power rifle
- ☐ Cleaning kit for rifle
- ☐ Handgun
- ☐ Ammunition for handgun, 200 rounds
- ☐ Holster for handgun
- ☐ Belt for handgun holster
- ☐ Cleaning kit for handgun
- ☐ Shotgun
- ☐ Ammunition for shotgun, 200 rounds
- ☐ .22-caliber rifle or handgun

- ☐ .22-caliber ammunition, 5,000 rounds
- ☐ Spare parts, including firing pins for all firearms
- ☐ Lubricating oils and greases for firearms
- ☐ Reloading equipment and supplies
- ☐ Knife
- ☐ Knife sharpener
- ☐ Pepper spray

Nuclear, biological, chemical equipment:

- ☐ Gas masks
- ☐ Spare filters for masks
- ☐ Water canteen with drinking tube for gas-mask use
- ☐ NBC suits
- ☐ Radiation-detection equipment
- ☐ Personal dosimeters and charger
- ☐ Potassium iodide, 4 ounces per person
- ☐ Chemical agent detection papers

Medical supplies:

- ☐ First-aid book
- ☐ Disposable rubber or latex gloves
- ☐ Painkillers such as aspirin or Tylenol
- ☐ Triangular bandages
- ☐ Pressure bandages
- ☐ Gauze bandages, 2" wide
- ☐ Q-tips
- ☐ Butterfly closures
- ☐ Antiseptic solution
- ☐ Sterile dressings, 4" × 4" square
- ☐ Adhesive dressings
- ☐ Nonadherent dressings
- ☐ Adhesive tape
- ☐ Absorbent cotton
- ☐ Mercurochrome

- ☐ Cortisone itch cream
- ☐ Oil of cloves (for toothache)
- ☐ Ipecac syrup
- ☐ Tincture of benzoin
- ☐ Antibiotic ointment
- ☐ Antibiotics (Doxycycline, Bactrim, etc.)
- ☐ Neosporin ophthalmic ointment
- ☐ Tinactin antifungal cream
- ☐ Antihistamines (Atarax)
- ☐ Pepto-Bismol
- ☐ Sudafed decongestant tablets
- ☐ Irrigation syringe with plastic tip
- ☐ Calamine lotion
- ☐ Safety pins
- ☐ Petroleum jelly
- ☐ Oral thermometer
- ☐ Small scissors (blunt ended)
- ☐ Tweezers
- ☐ Remover forceps
- ☐ Antibacterial hand wipes
- ☐ Baking soda
- ☐ Eyewash
- ☐ Oral rehydration packets
- ☐ Snake-bite kit (where applicable)

Transportation equipment:

- ☐ Mountain bike
- ☐ Bicycle pump
- ☐ Tire-patching kit
- ☐ Wheelbarrow or cart
- ☐ Sled or cargo toboggan
- ☐ Hand-operated fuel pump
- ☐ Tool kit for car (pliers, jumper cables, spotlight, screwdrivers, wire cutters, etc.)

☐ Skis or snowshoes in northern areas

Miscellaneous:

☐ Insecticides
☐ Insect repellents
☐ Solar charger or power generator
☐ Important personal papers

Barter supplies:

☐ .22-caliber ammunition
☐ Lighters and matches
☐ Contraceptives
☐ Liquor
☐ Cigarettes and tobacco products
☐ Coffee and tea
☐ Candy
☐ Feminine-hygiene supplies
☐ Silver and gold coins
☐ Fishhooks
☐ A trade or profession suitable for survival use

Guidelines for Equipment and Supplies

Food

Food is essential in all the scenarios. Your choice of provisions will play a large part in how you come through a situation. You should plan your rations. Have foods that are ready to eat, have some that take minimal preparation, and have some that are suitable for a base camp where you will have time to cook. Whether you use a camp

stove or open fire will also affect the type of food you are eating. If cooking on a stove, you will be limited to one-pot meals unless you have a lot of time. A campfire lets you heat more than one pot at a time and is much more suitable for baking.

The weight of the food is another important aspect of rationing, particularly on longer treks. Compare the calories against the weight of the food you intend to bring. Dried foods are usually the best as they are light and won't spoil in warm weather. Freeze-dried foods, which contain only 2 percent of the original moisture, are often lighter than dried foods, in which 25 percent of the moisture can remain. How food is packaged also affects weight.

The amount of food a person needs each day depends on a variety of factors. First, the intensity and amount of activity you are doing will be the major influence on how much fuel you burn. Another factor is age. An active, young person will burn food quicker than someone who is older or heavier. The height and weight of a person are other factors in the amount of food an individual needs. The outside temperature also influences fuel consumption as the body uses energy to keep warm on cold days. Due to these variables, it is difficult to recommend a precise amount of food. But as a rough figure, a person participating in outdoor activities will need between 2,500 and 3,500 calories per day. This works out to approximately two pounds of food.

The human body can exist for an amazingly long time on little or no food, but needs water regularly. Dehydration can cause heatstroke, hypothermia, frostbite, mountain sickness, and death. Drink water regularly, even when you don't feel thirsty. In the winter, you need at least three or four quarts each day. Add another quart in the summer. Keeping track of urine output is a good way to monitor your water level. You should urinate approximately three times a day. The liquid should be clear and light colored. A dark yellow color is a sign of dehydration.

Food consists of three main components: fats, carbohydrates, and proteins. All provide essential energy, but are assimilated and used differently. Vitamins and minerals are also essential to long-term survival, but short-term deficiencies are not likely to cause serious harm.

Carbohydrates are the food most quickly turned into energy and should thus compose the bulk of your diet, about 65 percent, when exercising. There are two types of carbohydrates: complex, which are starches such as grains and vegetables, and simple, which consist of sugars like sucrose and honey. Complex carbohydrates are the best fuel for strenuous activities because they provide more energy over a longer period. Simple carbohydrates give bursts of energy that burn off quickly.

Fats, which are present in foods like eggs, nuts, meat, and milk products, should compose about 25 percent of your daily calories. Increase this amount by about 10 percent in winter because fats have an important role in making the body tissues less sensitive to cold.

Protein, the raw material that renews the body's cells, is contained in food such as meat, beans, and nuts. Although protein is crucial to survival, it should compose only about 10 to 20 percent of your diet.

Modern prepared foods have excellent shelf lives. Some examples are shown below:

Canned food	2–3 years
Dehydrated foods	3–4 years
Freeze-dried foods (pouches)	2–4 years
Freeze-dried foods (cans)	10–15 years

Stoves

A portable stove is an indispensable tool for both backpackers and travelers. With a stove, hot food or purified water can be ready in minutes even under the most adverse conditions. Choosing the best stove for your needs can be a complicated task with the variety of fuel types and models available. To narrow the choices, consider the following questions:

- How lightweight and compact is the stove?
- What level of heat does it produce?
- Is appropriate fuel available where you are traveling?
- How safe is it to operate?
- Is the model reliable?
- Can the stove be field maintained?
- Does it simmer well?

It is necessary to know something about vaporization to understand how a stove works. Vaporization is the process whereby liquid or solid becomes gas and mixes with oxygen to become combustible. This process works differently with each fuel type. In gasoline and kerosene stoves, heat is used to expand the liquid fuel into vapor inside a vaporization tube. This process pushes the gas through a jet and toward a burner plate. Oxygen is sucked in along the way, and the fuel is then ready to light. In the case of liquid petroleum gas stoves, heat is not required to create vaporization because the gas is already a vapor when it leaves the canister.

The operation of a stove is affected by both altitude and wind. At higher altitudes, stoves have less oxygen available for combustion because air density is less. This causes water to boil at a lower temperature at high elevations, so food takes longer to cook.

Wind is another negative influence. It can decrease heat output by up to three times in some conditions. To prevent heat loss, erect a windscreen around the pot and burner to prevent warmth from blowing away. Be careful not to cover the fuel tank with the windscreen as the fuel tank may explode.

Gasoline—Gasoline stoves tend to be the most popular type for travelers. They are generally cheap to operate, come in lightweight models, and burn very efficiently. Also sold as naphtha, camp fuel, or Coleman fuel, white gas is widely available in most parts of the world. The problem comes during a disaster scenario. Once the fuel on hand is exhausted, your stove is just so much scrap metal.

Although white-gas stoves are understandably popular because they are easy to use and cook food quickly, they might be unsuitable for some users. The primary liability with white-gas stoves is that their fuel is very volatile. In addition, some stoves may require a fair amount of maintenance.

Multifuel stoves are usually white-gas stoves that can be adapted to other fuels. They offer users the option of running the stove on a second fuel if the first is not available. It is important to note that automotive gas should *not* be substituted for white gas as the additives put in vehicle fuel create lethal vapors and can clog the fuel lines.

Kerosene—Kerosene stoves have drawbacks and advantages similar to those of gasoline models. The primary benefit to using kerosene is the fuel's low volatility, which makes it less liable to explode or start a fire. It will burn hotly though, and travelers off the beaten path will find kerosene more readily available than most fuels. On the negative side, kerosene's strong odor will cling to belongings and eventually suffuse an entire pack. If using a kerosene stove, you will need a separate priming agent—such as white gas, alcohol, or priming paste—to ease vaporization.

Liquid Petroleum Gas—Liquid petroleum gas is a generic name for commonly used fuels such as butane, isobutane, and propane. Most LPG stoves are easy to use and require relatively little maintenance. The fuel comes in pressurized metal canisters that are common in most parts of the industrialized world.

Liquid petroleum gases are affected by temperature more than other fuels. When the temperature outside decreases, the pressure inside the fuel tank decreases as well. Consequently, most LP gases will not vaporize in colder climates. For instance, at sea level, isobutane will only vaporize at 15°F or warmer. Propane stoves are the exception as they will work down to −49°F. LP gases are also affected by altitude, which improves their ability to vaporize. Thus LP gas stoves are a good choice for travel at high elevations.

Methyl alcohol—These simply constructed stoves burn a wood derivative also known as methyl hydrate, alcohol, marine stove fuel, gasoline antifreeze, or methanol. Not available in some parts of the world, methyl alcohol is the only stove fuel that burns without pressure and has an invisible flame. Methyl alcohol does not burn very hotly. It will only produce half as much heat as the same weight of gasoline or kerosene.

Solid fuel—Solid fuel stoves use anything from fuel tablets to twigs. Thus fuel is available in the country. Stoves with blowers are better for cooking because of better heat distribution. The drawback is a blower needs batteries. Without batteries, solid fuel stoves are quite anemic. These stoves are essentially unaffected by altitude except for the longer cooking times required.

Sleeping bags

There was a time when buying a sleeping bag was simple—get the traditional duck pattern or opt for the ever-popular guns-and-fishing-poles design. Today, there is much more to consider:

- Comfort rating
- Type of fill
- Shape and construction
- Weight and compactability

First consider the typical conditions that you expect to encounter. Sleeping bags can generally be grouped into four comfort-rating categories:

- Summer weight (above freezing)
- Three-season (as low as 15°F)
- Four-season (5°F to 0°F)
- Winter/extreme (as low as −40°F)

Comfort ratings are based on the insulation's thickness, or loft (more loft equals more heat retention). Use these ratings as a guideline because there are several real-life factors that will affect the bag's performance. Your metabolism, diet, and fluid intake directly affect how warmly you sleep. Ground insulation, shelter, and the shape of the bag will also contribute dramatically to your overall comfort.

When fully lofted, the fill in a sleeping bag creates thousands of very small dead-air pockets. These pockets trap warmth generated by your body, slowing transfer to the hostile environment beyond the sleeping bag.

Waterfowl down and synthetic fiber are the materials of choice because they are durable, do not conduct heat very well, and have the capability to loft time after time.

Down, as any waterfowl will tell you, is an excellent insulator. Nothing beats the warmth-to-weight ratio, compressibility, or luxurious feel of a good down bag. These characteristics make it an ideal choice for those who travel light or want to minimize pack space. The quality of down is measured by its fill power; e.g., one ounce of 550 fill has the volume of 550 cubic inches when fully lofted; 500 to 550 is good. A 600 to 700 fill is excellent, but expensive. But, with the proper care, a down bag will last much longer than a synthetic fill bag, making it a good long-term investment. The major drawback with down is that it loses most of its insulating value when wet, and air-drying takes a very long time. Down may also be a problem for some allergy sufferers.

Synthetic fills are composed of small-diameter polymer fibers. Sheets, or batts, of these fibers provide the insulation used in many sleeping bags now available. Quality bags will contain respected brand-name fills such as Hollofil II, Polarguard HV, or Lite Loft.

Hollofil II is a soft, relatively compressible fill composed of short, hollow fibers. Polarguard HV (High Void) is made up of continuous, cross-linked, hollow-core polyester fibers. This ensures that the insulation stays where it is supposed to, making Polarguard HV the most durable synthetic fill. 3M's new Lite Loft, composed of extremely fine, cross-linked fibers, approaches the warmth-to-weight ratio of down. Lite Loft has proven to be durable, compressible, and comfortable.

Although not as compressible or as light, a synthetic bag is less expensive than an equivalent down bag. Also, synthetics do not absorb water. This means that, unlike wet down, a synthetic fill will maintain some of its lofting power and warmth when wet. Synthetics will also dry on their own faster than down, making them less vulnerable to mildew and other moisture damage.

Building a better bag—The outer shell must serve three critical functions:

- Hold the insulation in place
- Create an effective wind block
- Allow internal moisture (perspiration) to escape

Ripstop or taffeta nylon are often used because of their lightness and tight weave. A tight weave will keep the fill, especially down, in place and minimize wind penetration. Gore-Tex is used on some high-end down bags. Although it adds a bit of weight over regular nylon, Gore-Tex provides a great deal of moisture protection and wind resistance, making for a warmer bag.

For comfort and efficiency, the inner lining must be soft, breathable, and be able to wick moisture away from your body. Nylon meets these needs, but a nylon/poly/cotton blend will feel better against the skin.

To prevent the outer shell and inner lining of a synthetic bag from touching a potential cold spot, sheets or batts of fill should be overlapped so that each stitch line is backed up by one or more layers of insulation.

Because down is a loose fill, compartments are needed to keep it in place. Baffles, made of mesh fabric, are sewn from the shell to the lining to create channels that are then filled with down. Using baffles also prevents the shell and lining from touching, eliminating cold spots. Some lightweight bags are sewn through (the fill is stabilized by stitching shell and liner together). These bags are recommended for warm weather use only because each stitch line creates a cold spot.

There are four basic bag shapes: mummy, modified mummy, barrel, and rectangular, each with its own inherent advantages and disadvantages. With less space to warm, a close-fitting bag will have a higher thermal efficiency than that of a roomier one, even if they have the same amount of loft. The trade-off is that it may feel too constricting to be comfortable. Check the fit by climbing in and zipping up. Make sure you can move freely without compressing the loft, especially around the shoulders and in the foot box. If you don't feel comfortable, try another bag. It won't get better when you set up camp.

Hoods should be contoured with lots of insulation (up to 40 percent of personal heat loss is through the head).

Yokes will keep warm air from being forced out around your neck whenever there is movement in the bag.

Foot boxes should allow room for your feet to rest naturally without compressing insulation. Also look for additional insulation in this area.

Draft tubes should run the entire length of the zipper to prevent cold spots. It is best when they are sewn to the inner lining, not sewn through to the outer shell.

Zippers made of nylon are light, easy running, and conduct heat less than metal. Nylon coil zippers are less likely to snag fabric and are self-repairing if untracked.

Accessories

Liners made of soft poly/cotton blend can be used to keep the inside of the bag clean. They add some weight, but save your bag from the trauma of over-washing.

Overbags are simple synthetic-fill bags that slip over your regular bag to extend the comfort range. Alone, they can sometimes be used as a summer bag.

Sleeping pads provide ground insulation, protection, and added comfort. Consider one of closed-cell foam (won't absorb moisture) or a Therm-a-Rest (open-cell foam in a waterproof, airtight case). Unprotected open-cell foam acts like a sponge if it gets wet.

Care and cleaning—Store your bag loosely, never compressed in its stuff sack. Air it out, and make sure it's dry before putting it away in a cool, ventilated area (hung up or in a large storage sack).

Use a down-specific cleaner such as Kenyon or a very mild detergent when washing a bag. Soap leaves a residue even after repeated rinsing, and harsh detergents can deteriorate the fill. For both down and synthetics, hand washing in a large tub of luke-warm water is safest, but a commercial, front-loading washer will also work. Soak the bag, and then work the cleaner with a gentle agitating motion. Rinse repeatedly until the water is clear. Then press out (never wring) as much water as possible. Always support a wet bag entirely when picking it up. The weight of soaked fill can destroy internal baffles and stitching. Unlike synthetic fills, down will take a very long time to air-dry completely (three to five days), an invitation to mildew. If possible, use a commercial dryer at a low heat. To break up any clumps and regain the loft of down, put some clean tennis balls in the dryer.

Clothing

Dressing in layers, instead of one bulky do-everything garment, can help prevent uncomfortable and potentially dangerous situations. Layering is the relatively simple concept of dressing in a way that allows you to adjust to a wide range of environmental conditions.

To understand layering a little bit better, let's look at some of the mechanisms the body employs to regulate comfort, because after all, comfort is what it's all about. To provide an optimum working environment, your internal systems try to maintain a thin layer of warm (86 to 91°F), still air around your body. This is your very own microclimate. If the surrounding environment was constant and your life was void of activity, it's all you need. But that is not the case. Once you step outdoors, throwing caution (and your body) into the wind, you run the risk of knocking your microclimate out of whack. Physical activity, wind temperature, and moisture can all contribute to creating conditions too extreme for the body's mechanisms to deal with on their own. Wearing a series of thin layers will allow you to maintain an optimum microclimate during periods of physical exertion, as well as those times you stand around waiting for something to happen. By dressing this way, you can fine-tune your microclimate by shedding layers before you get too hot or by adding layers before you start cooling down.

Inner layer—Referred to by most people as underwear, this layer plays the most critical function of the system. It must transport moisture, usually in the form of sweat, away from the skin and disperse it to the next layer where it can evaporate. This process is often advertised as wicking.

Why is this so important? Water is a very good heat conductor. A wet garment against your skin can draw heat away from your body twenty-five times faster than a dry one. Even in conditions above freezing, this rapid heat loss can cause a dangerous drop in your body's core temperature, leading to hypothermia. Synthetics such as polypropylene and polyester now dominate as the materials of choice for this layer. Synthetics are light, strong, and best of all, unlike natural fibers, synthetics absorb very little water. This quality makes for quick-dry materials, reducing the risk of conductive heat loss.

Synthetic underwear is available in light, medium, and heavy weights to meet the demands of different activities. Light weight is for sustained activity where moisture transport is paramount. For changing activity levels, midweight allows a balance of wicking and insulation value. Heavy weight is for those times when your activity level is limited to watching the temperature drop. The inner layer should fit snugly, but not so tight as to feel restricting.

Midlayer—The prime directive, so to speak, of the midlayer is to provide insulation and continue the transport of moisture from the inner layer. To slow heat loss, this layer must be capable of retaining warmth that is generated by your body. Wool and synthetics are well suited for this purpose because the structure of the fibers create small airspaces that trap molecules of warm air.

As far as moisture management goes, synthetics have the upper hand because they absorb little water, allowing faster evaporation. Wool absorbs up to 30 percent of its own weight in water, leaving it heavy and difficult to dry. Synthetic fleece/pile garments (jackets, pullovers, and vests), as well as being lightweight, are very durable and require less care than wool. Additional features such

as pit zippers and full-length front zippers add to the versatility and control. As with the inner layer, this layer should be snug, but not constricting. If it is too loose, it just means more space your body has to warm up.

Outer layer—With all the effort you have just put into creating the perfect environment, you don't want to have it blown away by a gust of wind or soaked by rain. The outer layer protects your micro-climate from the elements. It should also allow air to circulate and excess moisture to escape.

Choose on the basis of what you plan to do, where you plan to do it, and what you plan to spend. For dry conditions, a breathable (uncoated) wind shell may be all you need. If you expect conditions to be more severe, a waterproof (coated) rain jacket with additional insulation may be more in line with your needs. A shell made of a breathable/waterproof fabric, such as Gore-Tex, will give you protection from wind and rain, as well as allow water to escape. Keep in mind, however, there are no miracle fabrics. When you are active, your body can produce more water vapor than any fabric can pass. The result can be a buildup of moisture on the inside, leaving you wet, clammy, and cold. Strip off a layer or open any ventilation zippers before this happens, and you will be a happier camper.

Head protection—It has been estimated that 40 percent of a person's total heat loss can occur through the head. This is because your body considers the head to be a rather important extremity and therefore pumps a hefty volume of blood to it, keeping it warm and functional.

Unfortunately, contrary to what some people may say, the head does not have much fat. Without this natural insulation, your head acts like a radiator, letting heat escape. This puts a strain on the rest of your system because your body must now use additional energy to rewarm the blood as it recirculates. A good wool or fleece hat will not only slow heat loss through your head, it will also make your hands and feet feel warmer because of the improved circulation. Don't overlook full-face balaclavas and neck gaiters for those really harsh conditions.

Hands and feet—In its effort to keep your head and torso warm in cold conditions, your body reduces blood flow to the hands and feet. To compound this, these areas do not generate much heat on their own. Some sort of protection is needed.

Mittens are warmer than an equivalent pair of gloves because the whole hand contributes to the warming process. The disadvantage is a sacrifice in dexterity. Gloves are good for activities that require independent finger control, such as tying knots, but each finger must warm up its own little compartment, making them less efficient at keeping your hands warm. A layering system that consists of a thin wool or synthetic glove for moisture transport, an insulating mitten, and a noninsulated shell mitten for outer protection will give you a wide range of temperature control and manual dexterity.

Wool—Wool is still the dominant choice for socks, offering a good balance of moisture management, insulation, and cushioning. The addition of a polypro liner sock will speed up moisture transport from the feet to the outer wool layer. Socks should fit snugly; wear them too tight, and circulation can be restricted. A loose sock can slip or bunch up, creating pressure spots that can lead to blisters.

Gore-Tex—Gore-Tex is actually a membrane that is bonded to other high-performance materials to create a durable, laminated fabric that is waterproof and functionally breathable. The Gore-Tex membrane is made up of material (expanded polytetrafluoroethylene) that has about nine billion microscopic pores per square inch. These pores are 20,000 times smaller than a single droplet of liquid water, but 700 times larger than a molecule of water vapor. This allows water vapor from perspiration to easily pass through the membrane while keeping liquid water from the outside at bay. Integrated into the membrane is an oil-hating substance that prevents contamination from insect repellents, cosmetics, or misguided chunks of that greasy chili you had for lunch.

Gore-Tex is very stable. It can withstand just about any conditions that you could encounter on earth or, in some cases, beyond. It remains functional at extreme temperatures. It is not damaged by saltwater or ultraviolet light. Bleach, detergent, and dry-cleaning chemicals will not harm it, and it is not susceptible to mold or mildew. Just remember that the other fabrics used in the garment may be affected by these agents. The care tag included with the garment will have more information.

The value of keeping your Gore-Tex clean cannot be overstated. Keeping water beading on the surface of the fabric allows the membrane to breathe better and minimizes condensation, resulting in vastly improved comfort. Dirt can dramatically impede performance so it is a good idea to wash your Gore-Tex garment frequently, following care instructions on the sewn-in label.

The material is very durable, so don't worry about damaging the garment with vigorous treatment. (However, double-check the care tag for special handling if the garment uses silk, wool, or down.) A second rinse cycle may be useful in removing detergent residues. High heat in the dryer and even a light touch-up with a warm iron will help restore water repellency. After extended wear, use a commercially available water-repellent spray.

Have spare clothing on hand. This is desirable as, under some circumstances, contaminated clothing should be discarded. You really don't want to cavort around naked during a disaster.

Boots

Selection of footwear is as important as selecting the right clothes. Many would argue that it is even more important. Depending on the need, we have hiking boots, mountaineering boots, and snow boots. Boots are further divided into cold-wet and cold-dry types. Given the great variety of boots on the market, we shall limit discussion to a few types.

For all-around hiking, insulated, Gore-Tex hiking boots are recommended. The plastic, double boots made by Koflach or Scarpa are the top of the line for mountaineering. For the snow, the Canadian Sorel boots have rubber bottoms triple-stitched to leather uppers and come with liners.

Heavy mountaineering boots are not designed for walking, so stay with hiking boots for all-around use. Don't buy boots that are

too tight. Your feet will swell when you are hiking. You buy boots to protect the feet from the roughness of the trail and to keep out water, snow, mud, stones, and other material. Any boot you buy should have a lugged outsole. Some of the better ones are Vibram Foura and Vibram Hiking. The boots should flex at the balls of the feet and be made of full-grain leather with a Gore-Tex bootie. This is reasonably watertight.

Boots higher than the ankle add weight and tend to form creases that irritate the Achilles tendons. Truly waterproof boots are for specialized use only. The feet sweat and the boot thus becomes wet anyway.

Boot care—Wear new boots around for a couple of weeks, and let them get scuffed up. This helps to remove the factory-applied buffing wax. Then treat the seams and the join of the sole with Free-sole or a similar compound. This will greatly increase the water resistance and durability of boots. Allow the treated boots to cure for twenty-four hours by placing them in a box with a wet towel. This treatment should be done before applying other water-repellent compounds.

Treat the fabric and leather with silicone-based compounds, which will provide excellent water resistance. Using a toothbrush, apply a thin coat and allow twenty-four hours for it to penetrate the leather and fabric before using.

Water treatment

All types of water are susceptible to contamination—from desert springs to alpine streams—and there is no way to tell if water is pure or not. Consequently, prudent travelers will want to disinfect any untreated water.

In the wilderness, the primary threat from untreated water is the protozoa *Giardia lamblia*, which is transmitted from the feces or cysts of humans and animals. Often called beaver fever, giardiasis can cause cramps, diarrhea, nausea, and vomiting. Travelers in countries without water treatment must protect themselves from giardiasis as well as a variety of viruses and bacteria that are present in human sewage. These pathogens may cause diseases such as typhoid, cholera, hepatitis, and dysentery.

Whichever method of disinfection you choose, always use water that is clear and as fresh as possible. To ensure water does not contain toxic amounts of salts or minerals, look for a source that is home to aquatic life and surrounded by plants. After finding the purest water available, use one of the following methods of disinfection:

Boiling—An effective and simple method of water treatment, boiling can destroy the majority of parasites, viruses, and bacteria. To kill most microorganisms, including *Giardia lamblia*, you will need to boil water for approximately two to three minutes at sea level. The boiling time will increase at higher elevations because water boils at lower temperatures. For instance, at 5,000 feet, water needs to boil for about four minutes. If traveling higher than 5,000 feet, add five minutes of boiling time for each additional 5,000 feet of elevation gain.

There are several drawbacks to using boiling as the method of water treatment that make it unsuitable for some types of travel. The major liability is that boiling takes up fuel and time, particularly at higher elevations. Moreover, there is a waiting period until the water becomes cool enough to drink, debris is not removed, and the water is left with a flat taste.

Chemical disinfection—Iodine is the most popular chemical for use in backcountry water disinfection. Adding it to water in the form of tablets, tincture, or crystals, iodine is lightweight, cheap, and takes up little room in a pack. On the negative side, it can have adverse effects if ingested in large quantities or over a long period. Consequently, iodine should only be used as a temporary method of water disinfection. Any person that is pregnant or suffering from thyroid problems should avoid using it at all.

If used properly, iodine will kill bacteria, viruses, and protozoa. Using iodine *properly* can be difficult. The problem is that the amount needed varies depending on the water's temperature and

pH as well as the amount and variety of matter present. You will need to use enough to kill any dangerous microorganisms without poisoning yourself. Of particular concern is iodine's uncertain effect on *Giardia*. Manufacturers often claim that iodine is 100 percent effective in killing *Giardia*, but these statements usually are based on tests performed in ideal situations. Some research has suggested that there are too many unknown variables to make iodine treatment completely reliable.

If you're using iodine, there are some procedures that will increase your chances of killing any nasties. The most important thing is to follow the manufacturer's directions on how to use their product. Before treating, remove solid materials from the water because pathogens attach to matter. Do this by letting the water settle before treating and/or filtering it through layers of cloth. The longer iodine contacts the water, the better chance it has to kill microorganisms. It also works better in hotter water. So let cold water warm before treating.

Always keep iodine in a sealed bottle. It can cause a poisonous level of toxic fumes if left to air inside an enclosed space like a tent, and it can oxidize a number of materials including metal and cloth. Opened iodine tablets have a limited short life. Remember to replace them each year.

Filtering—Water filters are popular with many travelers due to their low weight, ease of use, and effectiveness in removing debris and microorganisms. Whereas boiling and chemical disinfection destroy microorganisms, filters attempt to remove them from the water altogether. This is accomplished by pumping water through a filter that lets water pass, but blocks pathogens.

The materials used in filters differ, but can all be judged by the size of microorganisms—which are measured in microns—they can remove from the water. Most good filters have screens rated to at least 0.2 microns, which will remove most pathogens and bacteria. A filter designed to screen only larger microorganisms, such as *Giardia*, need only screen to 1.0 microns. No water filter is designed to screen viruses as they are smaller than the smallest filter made. If traveling where viruses are a concern, treat water with iodine.

When choosing a water filter, be aware that cost usually reflects reliability and durability. In other words, good ones are expensive. An efficient unit will have a removable filter that can be cleaned and replaced when necessary. The longer the filters last before needing replacement, the more economical the system will be to use. Other features to look for are quick filtration and a high output. The filter should also be solidly designed with sturdy gaskets that will prevent leakage of untreated water into treated water. Ceramic filters when dropped can crack and become useless.

Water desalination—You can produce fresh drinking water from seawater with desalination units using reverse osmosis. They are expensive, but reject 98 percent of the salt.

Tents

A tent is one of the more essential and expensive items needed for sleeping outdoors. With a lightweight tent, you can carry a home on your back that can be assembled in minutes. No more lugging around heavy yurts or tepees. Tents are made in a multitude of designs to suit a wide variety of uses: from mountaineering to cycle touring to car camping. But a tent that is perfect for one use is sometimes not the best type for another. Therefore, you will probably want to find a tent, or tents, that work well for different activities. To narrow your choices, consider some basic questions. When and where will it be used? How many people must it hold? Is weight important? How much will it cost?

Three-season, four-season, and expedition tents—The first thing to decide is what time of the year you will be using the tent. Three-season tents can be used in the more moderate weather of spring, summer, and fall, while four-season tents can take you into winter. Four-season expedition tents go past the four-season rating and are designed to stand up to the worst possible conditions.

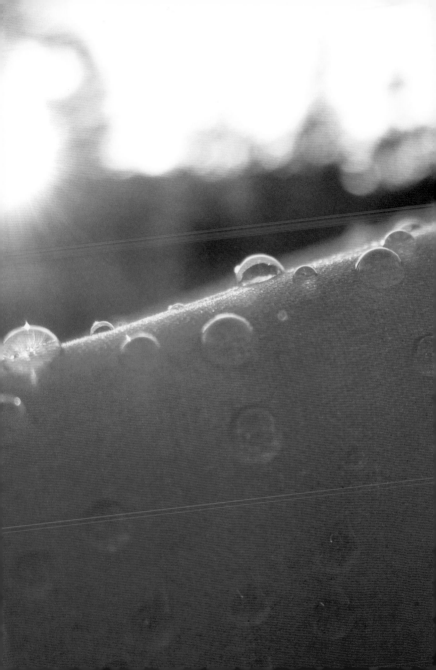

Tents designed for really bad weather usually feature more and stronger poles, additional waterproofing, and a sturdier design. Be aware that the season designation of a tent is not a perfect classification. A tent that stands up to snow well may not hold up in a really strong wind and vice versa.

Tent shape and design—The majority of tents have a breathable inner canopy and a separate waterproof fly. The two-walled construction is beneficial in two ways. First, it moderates the temperature, thanks to the insulating airspace between the two walls. Second, it allows ventilation of body vapor. Ventilation is important because a person loses about a pint of water during a night's sleep. If not allowed to escape, this moisture condenses inside the tent and can get you and your gear wet. A tent can also be ventilated by using waterproof and breathable fabric similar to Gore-Tex.

Traditionally, tents were made in an A-frame design that is becoming obsolete due to the evolution of flexible tent poles. Tunnel and geodesic dome tents, which use flexible poles, are now standard. Most tents are hybrids of these designs, and one manufacturer's version of a dome tent can be quite different than another's. This makes generalizations about design difficult.

Many tents are freestanding, which means they come to life without being pegged down. This is advantageous because the tent can be assembled anywhere and then plopped down in the best site. A freestanding tent works best when pegged and must be anchored in winds or when left unattended. Don't let freestanding be too important a criterion as some nonfreestanding tents may only require a few pegs and will be lighter and tighter than a similar freestanding model.

What to look for—Before buying a tent, you should set it up. If the tent is already assembled, ask the salesperson if you can take it down and put it up again. This is the best way to judge how easy it is to pitch. While setting up the tent, imagine you are cold, hungry, and wet. How long does it take? Could you do it wearing gloves? Once you have finished the assembly, take your shoes off,

get inside, stretch out, and roll around. Consider how many people will fit inside. Are the planned occupants particularly small or large? Is there room for gear?

Next, take a look at the inside of the tent. To determine how strong it is, push on the walls from different angles and directions to ensure it won't collapse or bend too easily. (Freestanding and pegged-out tents will behave differently in this test.) The canopy and fly should be as taut as possible since loose panels will decrease breathability, water repellency, and stability. Check that the poles join easily and snugly with no area of unusual stress. In general, tents with more poles and pole intersections are stronger, but heavier. Steeper walls catch more wind, but shed rain better. A flat roof allows snow buildup.

Tent construction—the body—A tent's floor should be made from a tough, waterproof material that is also used on the first six inches of the tent walls. The rest of the tent body will be constructed of a nonwaterproof fabric, usually taffeta or ripstop nylon. The weight of the fabric will largely determine how heavy the structure is, but keep in mind that lighter fabrics are usually less durable.

It is important to examine a tent's seams as they are prone to leaking and take stress when the tent is taut. Two seam types, bound and lap felled, are most common in tents. The weaker of the two, bound seams are constructed by stitching through a layer of material folded over the two pieces being joined. With lap-felled seams, the two pieces of fabric are placed on top of each other, folded, and then stitched. Good quality tents use lap-felled seams wherever possible and bound seams where more than two pieces of fabric need to be joined (although they should not be used indiscriminately).

Poles—Most tent poles are now shock corded, which means the poles are hollow and attached by an elastic cord running down the center and fixed at both ends. This makes the poles easier to assemble. Ideally, a tent pole should be lightweight, strong, and somewhat flexible. The three basic materials used in tent-pole construction are fiberglass, carbon fiber/composite, and aluminum.

Aluminum, the most common material, comes in a variety of qualities made from different alloys. The 7000 series is the strongest followed by the 6000 and 2000 series. Carbon fiber/composite poles are also popular because they are light and strong. But they are expensive. At the other end of the scale are fiberglass poles, which are cheap, heavy, and prone to breaking.

The fly—The fly is the part of the tent most exposed to the elements, and it will need to be replaced more often than the tent body itself. To be effective, a fly must be constructed of a waterproof material. It can be extended away from the tent to create a vestibule, which acts like a porch. Vestibules are useful for storing gear.

Stakes and zippers—Zippers are often the first part of a tent to break. If a slider goes first, it is relatively easy to fix, whereas an entire zipper is more expensive. When buying a tent, look for any tight or difficult zippers that could break later on. A nylon coil zipper is lighter and less likely to stick than a metal one. Aluminum stakes around nine inches long work well in most environments. Carry them in a separate bag to protect the tent from punctures.

Tent care—Proper care of a tent, both at home and in the field, can lengthen its life. Clean your tent after use as debris can create holes and tears once the tent is packed. When possible, let the tent dry before packing because moisture causes mildew that can stain the fabric and make it smell. Make sure it is completely dry before storing it for any length of time. In addition, seal your tent regularly (if not seam taped) because water tends to leak through the stitch holes. Lastly, if it is set up for long periods, cover your tent with a tarp or keep it in the shade to guard against ultraviolet rays, which are the single greatest cause of tent failure.

Binoculars

Binoculars are handy for many types of activities. They can be used for viewing wildlife, route finding, or peeking into farmhouses en route. The focus here will be on portable, lightweight binoculars, how they work, the different kinds, and what to look for when buying them.

Your eyes see things by reacting to light that has bounced off an object and traveled to them. Binoculars make distant objects nearer by passing this light through a series of magnifying lenses. The route the light takes through the binoculars is called the optical path. The longer the optical path, the larger the image that appears to your eyes. However, when you simply lengthen the optical path, the binoculars become big and bulky. The way around this, which is used in most modern binoculars, is to place a set of prisms between the lenses. The prisms make the light do several turnarounds to lengthen the optical path while minimizing the binoculars' size.

Types of binoculars—Binoculars can be classified by the way their lenses and prisms are arranged. The simplest design is the Galilean binocular, which is essentially two telescope tubes placed side by side. There are no prisms in the optical path. Prism binoculars come in two types: roof prism and Porro prism. With roof prism binoculars, the lenses sit in a straight line. The two lenses are offset in the Porro prism design. Either design can be used to construct quality binoculars, but Porro prism binoculars are usually cheaper, bulkier, and heavier.

What the numbers mean—If you look at any binocular, you will see numbers stamped on its body, such as 8×24, 7.10°. Respectively, the numbers represent the binoculars' magnification power, objective lens diameter, and angles of field of view.

Magnification—Magnification is the ratio of a subject's real size to its magnified size. This means a binocular with an eight-power magnification, usually written as 8×, makes an object appear eight times larger than it is. The majority of hand-held, general-purpose binoculars have powers of magnification that range between 6× to 10×. Most people find 10× the highest magnification they can hold because the vibration of an unsteady hand blurs the image in more powerful binoculars.

Objective lens diameter—The size of the objective lens partly determines the amount of the light that enters a binocular. The more light that enters, brighter the image is. A binocular's brightness also depends on magnification and lens quality, so you can't say absolutely that a larger objective diameter means a brighter image.

Angle of field of view—The width of the image you see through a binocular is the field of view. When you look through a binocular, the view widens, and you actually see a cone-shaped image. The angle of field of view is the angle of this cone. The larger the angle, the quicker your view widens as it leaves the binocular. A wide field of view is beneficial for watching large or moving objects.

Brightness—How bright an image appears through a binocular largely depends on its magnification power and the diameter of the objective lens. Independently, a larger objective lens creates a brighter image, while a higher magnification tends to make a darker image. One indication of a binocular's brightness is the size of the exit pupil, which is the bright circle seen on the surface of the viewing lens. You can determine the exit pupil's size by dividing the numeric value of the objective lens diameter by the magnification number. For instance, an 8×24 binocular will have an exit pupil diameter of 3 mm.

If you want to compare the brightness of different binoculars, you need to do more math. This is because the size of the exit pupil does not give a proportionate indication of how binoculars perform in relation to each other. To determine relative brightness, you need to square the size of the exit pupil. For example, an exit pupil of 3 mm has a relative brightness of 9, while one of 4 mm has a relative brightness of 16.

For night use, the exit pupil diameter should be at least 5 mm. Thus an 8×40 binocular or 7×35 are on the threshold of effective night viewing.

Lens coating—When light passes through a lens, some of it reflects off the surface and is lost. This reflected light causes a perceived reduction in an image's clarity, brightness, and contrast. A lens coating is a chemical applied to the surface of a lens or prism to decrease the amount of reflected light. Lenses can be single coated or multicoated. A single-coated lens has one magnesium fluoride coating. A multicoated lens has additional layers of other chemicals that further reduce reflection. You will also see the terms "coated optics" and "fully coated optics" to indicate how many of the lenses are coated. With fully coated optics, all lenses and prisms have coatings. A binocular with coated optics will only have treatment on some of the lenses and prisms. Ruby-coated objective lenses further improve contrast.

Sunglasses

Chances are, unless you're a hairy-tailed mole or a vampire, you spend a good deal of time exposing delicate body parts to the sun's potentially damaging radiation. Of course, common sense dictates that you take the appropriate precautions as far as sunblock and proper clothing are concerned. Likewise, effective eye protection should not be taken lightly. The wrong pair of sunglasses can fool you into thinking that your eyes are safe.

All light is not created equal—It's the invisible wavelengths at either end of the visible light spectrum—ultraviolet and infrared— that are potentially damaging to your eyes.

Ultraviolet light comes in three flavors:

+   UV-C is filtered out by earth's atmosphere.
+   UV-B is the nastiest, more so at higher altitudes or when reflected from water or snow.
+   UV-A is considered less harmful, but still presents certain risks.

Extended UV exposure can result in eye fatigue and a feeling of dry grittiness. It is also responsible for a temporary, though very painful, condition called snow blindness (sunburn of the eye's outer surface). And there is the possibility that long-term UV exposure may contribute to some types of cataracts.

UV inhibitors are incorporated into the lens material, be it glass or plastic, but it's impossible to detect them simply by appearance.

Check the manufacturer's specs or inquire about how much UV the lenses absorb.

Infrared radiation is felt as heat and can be easily filtered out or reflected with the proper lenses. Although it has not been proven to contribute to any specific eye damage, it may cause your eyes to feel dry or fatigued.

Many manufacturers now produce sunglasses that offer close to 100 percent ultraviolet and infrared protection. By the way, wearing untreated sunglasses (particularly glass) can be worse than wearing none at all. The pupil of the eye opens wider when shielded by the dark lens, allowing even more harmful rays to enter.

Transmission, coatings, and color—light transmission—This refers to the amount of visible light that actually reaches the eye. Although visible light is less damaging than light you can't see, it can still leave you squinting on a bright day. The light transmission, or darkness, of the glasses you choose should depend on how and where they will be used. Light transmission of around 15 percent is fine for bright, reflective conditions. For use around snow, water, or at altitude, 10 percent or less may be needed. Of course, these are just guidelines. The best way to determine what you are comfortable with is by trying different densities. Remember, transmission figures have nothing to do with the amount of UV being absorbed.

Lens coatings can be grouped into two basic categories, reflective and non-reflective.

Reflective coatings are mirrorlike in appearance and applied to the outer surface of the lens to reduce glare and heat (infrared). Traditionally silver colored, these *flash* coatings now abound in several funky colors, often named after some, as yet, undiscovered alien mineral or rare earth alloy. Reflective coatings are also available in different densities that vary the amount of light transmission through the lens, e.g., a gradient coating is darker at the top than the bottom. A double gradient is darker at the top and bottom, effectively cutting light from bright sky and ground reflections.

Nonreflective coatings are applied on the inner surface of the lens to absorb specific wavelengths and stray reflected light. These single or multilayer coatings are delicate. Follow cleaning recommendations carefully.

Lens color not only affects how cool you look to the world, but how the world looks to you. Color should be selected on the basis of the conditions under which they will be used. In general, gray or green lenses will give the truest color rendition, but at the cost of lower contrast. Amber lenses offer the best visibility under overcast, hazy, and low-contrast conditions. But some people don't like the color shift. You might consider gray-brown as a good compromise.

Special purpose lenses—Photochromatic lenses, available in glass or plastic, actually lighten or darken with the amount of UV light to which they are exposed. Unless the initial light transmission is 25 percent or less, these are not recommended as driving glasses because a car windshield filters out most of the UV that the lens needs to be activated.

Polarization—Polarization is possible through the use of specially treated plastics. Polarized lenses absorb light reflected from flat, shiny surfaces. This can be a big advantage, particularly around water or snow.

Glass or plastic lenses?—There are two schools of thought when it comes to lens material. It is easier to obtain high-quality, distortion-free lenses with glass, and they are much more resistant to scratching. The trade-offs are that glass is heavier than plastic and, even if chemically or heat tempered, prone to shattering if impact is sufficient.

High-carbon plastics, such as Lexan or CR-39, are now the material of choice with many manufacturers because optical quality has improved. They're lighter and generally less expensive than equivalent glass lenses. Although some plastics are stronger and highly resistant to impact, they scratch quite easily compared to glass and are heat sensitive. (Keep this in mind if you leave them on the car dashboard on a hot summer day.)

To check the optical quality of a glass or plastic lens, hold the glasses at arm's length, and look at a light source through them (fluorescent tubes are good). The image should be clear and undis-turbed through every part of the lens.

Functional frames—Among frames, nylon is safer, lighter, and more durable than metal or acetate. Metal frames can bend and could literally freeze to your face on cold days. Acetate is much more brittle than nylon, especially when cold. Whichever you choose, look for tough hinges and a comfortable fit.

Frames should not impair your peripheral vision. Also make sure they hold the lenses close enough to your face to keep out wind, bugs, and other airborne particles, yet allow air to circulate so fogging isn't a problem.

Care and cleaning—When not in use, sunglasses rank very high on the world's most-abused objects list. Get a case (hard plastic with padding is best), and use it. Consider retainer straps, especially when involved in water activities. They could save you the cost of new glasses.

When cleaning, remember that the coatings are delicate. Spit and pant leg just won't cut it. Use a cleaning solution when possible and lens tissue or a soft, lint-free cloth.

Rifle

Your primary tool for protection and hunting is your rifle. Military and military look-alike rifles are more rugged than civilian hunting firearms. This makes the military rifle superior in a disaster scenario. To give you an example, imagine walking through a forest then crossing a stream and then scrambling down a steep hillside. Note I did not say anything about firing your rifle. If you put your rifle aside after such a trek, you may find rust on many parts of it. It is a tribute to modern gun manufacturing on how well firearms stand up to abuse. A military rifle with its rust-resistant finish is better able to take the above treatment and keep on functioning. What is the criteria for selecting your rifle?

1. The rifle must be capable of consistent and repetitive one-shot kills on medium- or large-size animals.

2. The rifle must be completely reliable in its function and operation under any and all circumstances.

3. The rifle must possess the capability of delivering a high rate of fire in the event of multiple hostile targets presenting themselves.

4. The size of the individual round of ammunition should be as small as possible while meeting the stated killing requirement. The ammunition should be widely available.

5. The rifle must be sufficiently rugged to perform at peak efficiency for extended periods with minimal maintenance.

These are the main requirements. There are secondary requirements such as the ability of reloading the ammunition, the ability to install telescopic sights, the option of using a subcaliber device, and so on. Keep in mind that there is no rifle that combines all primary and secondary requirements into one, contrary to claims made to that effect. Subcaliber devices enable you to fire .32-ACP pistol ammunition in .30-caliber rifles and .22-rimfire ammunition in .223 Remington caliber rifles.

For defensive use, you will be best served by a semiautomatic rifle in .308, .30–0, or 7.62 × 39 caliber. The 5.56-mm NATO or .223 Remington may not be a reliable game-getter in spite of its current military use. The recommended, currently available rifles in the commercial market are: M1A1 Springfield Armory, M1 Garand, and the SKS carbine. The AR-15 family of rifles is chambered for the 5.56-mm NATO round and is considered adequate.

Handguns

Handguns are generally classed as defensive and working handguns. We will not cover so-called offensive handguns. The skill required to use them in that role requires almost single-minded application to practice, which is not available to most of us. The handgun should always be with you or no more than an arm's length from you. Whether you are in the latrine, making love, chopping

THE SURVIVALIST'S HANDBOOK

down a tree, your handgun should be on your belt or close to you. There are times when your rifle is in camp while you are working. Under attack, you will have to make do with a handgun.

There are two major classes of handguns—pistols and revolvers. A pistol can be rapidly reloaded with a fresh magazine and as such is considered to be a defensive handgun. Revolvers are able to feed and fire a wider variety of ammunition, but are much slower to reload. They are considered a working handgun. For example, if you are traveling in the desert, you want the first round in the chamber to contain pellets to kill snakes, scorpions, and the like. Most semiautomatic pistols will have a hard time feeding shot-shell loads, while revolvers have no such problems. On the other hand, should the local chapter of the Hell's Angels take an exception to your hairstyle and burst in unexpectedly while your trusty riot gun is in another room, a pistol with a high-capacity magazine will mean the difference between living and dying.

Most off-the-shelf handguns of reputable manufacturers are adequate for the job. Avoid customizing your handguns. Many otherwise sane people read about the latest custom changes available for handguns and go for them. They find that about as exciting as having a new mistress. What you need in addition to the handgun is a good holster, spare magazines, a good cleaning kit, and some spare parts. That is all.

What is critical for any caliber chosen is that it delivers instant stopping power. There are two schools of thought on this issue. First there is the old school of low-velocity, heavy bullet. Opposite is the high-velocity, lightweight bullet. Their arguments will continue long after firearms are obsolete. Rather than rehashing the arguments, let me say that the following calibers are adequate for use in defensive handguns: .357 Magnum, 9-mm Luger or Parabellum, .40 S&W, .44 Special, and the .45 Auto. I left out the 10-mm and the .44 Magnum. While they are very good one-shot man stoppers, they are large handguns handling with the grace of a sack of marbles and requiring more time for follow-up shots.

Recommended handguns are Colt or Springfield .45 1911A1 clones, Firestar pistols, Ruger and Smith & Wesson revolvers, SIG pistols, Browning pistols, Beretta model 92 pistols and its Taurus look-alikes.

The Shotgun

In the military, engaging of multiple targets close to you, sometimes at night, is achieved with fully automatic weapons. For the civilian, a good shotgun performs the same purpose. It is also useful on

moving targets. The variety of shotgun ammunition on the market-place makes the shotgun a very useful firearm. You can destroy doors, deliver tear gas, and blast through barricades with some of the special ammunition.

For defensive use, good short-barreled (18 to 20 inch) pump-action shotguns are preferred by most knowledgeable people. The problem with a pump-action shotgun is "short stroking" by a person under pressure. This occurs when the forearm is not pulled back far enough to pick up and load another shot shell. Another problem is a sling interfering with the operation of the action. The shotgun is strictly a short-range proposition, 70 yards or so. However, a properly sighted shotgun can deliver slugs out to 150 yards.

Rimfire rifles and handguns

The ubiquitous .22 round has killed game ranging up to a circus elephant. What it does not have is stopping power (which is the ability to stop an opponent immediately before he has the chance to harm you—the circus elephant dying a week after you shoot him is little comfort if he tramples you to death). If it had stopping power, we would not need the plethora of calibers currently on the market. Where the .22 excels is at pest control, hunting where the noise of a center-fire rifle may bring unwanted attention to you, and hunting small game.

The rimfire rifles or handguns can be had in many different configurations, all the way from bolt actions to fully automatic. Given the low power of the round, most people prefer to have semi-automatic firearms. If you are knowledgeable and have a chance to practice with a pistol, .22-caliber rimfire is perfectly adequate. This reduces the amount of hardware you have to carry with you. Rimfire ammunition may become universal currency in a prolonged survival scenario.

This section deals with saving your *Assets*. To start your thought process regarding economic survival, let us, by way of example, look at California. For millions, California has been the symbol of hope, wealth, and the American dream. *Go west, young man, and seek your fortune.* But no more. California is suffering from an economic depression. Military cuts have hurt its biggest industries. For the first time in 200 years, people are leaving California! To top it off, the crash of Japan will make this even worse.

A few decades ago, Japanese investors were instrumental in pushing up real estate prices in California. They spent billions of dollars. They could afford to—they were rich. However, during the 1990's, the Japanese have lost several *trillion* dollars of wealth. The Tokyo stock market crashed, losing 60 percent of its value, and as of 1997 has yet to recover. Recession is starting in Japan. Right now the Japanese are pulling out of California. And they are doing so in a big way.

Like the crash in Japan, the current economic crisis was caused by the global housing bubble. You can really get taken to the cleaners if you have invested heavily in any of the thousands of overpriced mutual funds. Lately new funds crop up like weeds in an empty lot. There are more mutual funds listed than stocks on the New York Stock Exchange—more baskets than eggs. Most fund managers think that's great, but inevitably it will result in a mutual fund disaster.

Economic Survival

It's a case of simple economics. As long as investors keep buying, most mutual funds will do *okay*. However, when interest rates go up, people will stop investing in mutual funds, and that will trigger a market blowup that will rival the crash of '87 and the bank-run panic of the Great Depression. Why? Because stock prices will fall as soon as fund managers stop buying. Panicky investors will want to pull out their money. Fund managers will have to dump stock to pay them off. It's a cycle that will feed on itself.

In many ways, it is similar to the savings-and-loan debacle. They were so heavily oriented toward real estate investments that when there was a softening in that market, they fell like dominoes. Today we see the banks heavily into mutual fund loans—a scary scenario.

Similarly, investing in government bonds is great as long as there is a government to redeem them. Investing in stocks is fine providing you know what you are doing. It is even better if the company you invest in is located close to you and you know something about the business. Depending on the size of your assets, you should invest them in the following way:

Type of asset	Under $50,000	Over $50,000
Barter goods	30%	10%
Gold and silver	10%	25%
Bank deposits	30%	10%
Stocks and Bonds	30%	25%
Real estate	—	30%

The table above is for general guidance only. You will have to prepare your own table. Analyze what you can do, and then decide whether you will become a trader, a scavenger, a farmhand, or whatever. Make your asset distribution table fit your abilities and resources. After doing it a couple times, you will get the hang of it. The asset distribution table is not carved in stone. Change it according to the scenario you are facing and as events unfold.

What not to buy:

- **Collectibles**—Their value depends on the number of people collecting. During hard times, people sell off things rather than buy. This applies to numismatic coins, proof sets, paintings, and vintage cars equally. The only exception was stamps during the Depression years.
- **Mutual fund**—Stocks and the like.

During inflationary times, it is better to buy real estate, while in a depression, it is better to rent.

Governments have the nasty habit of declaring the possession of gold illegal. It even happened in the U.S. after the crash of '29. Then the governments will declare an amnesty period during which you can exchange your gold coins for paper money. This will cost the government nothing more than running the printing presses overtime to gain really valuable resources.

Don't fall for this. Hide your assets. Do not keep them in a safety deposit box. Your silver and gold will be your stake in reestablishing yourself in the future. Eventually the government will introduce a new currency. In the meantime, you can ride out the rough times with your *hard* reserves. Although many countries hold reserves of gold against their currencies, no currency is fully backed by gold. The last currency fully backed by gold was the Swiss Franc but that ended in 2000.

Remember what Bank of England's Sir Josiah Stamp said: "Banking was conceived in iniquity and born in sin. Bankers own the earth. Take it away from them, but leave them the power to create money, and with a flick of a pen, they will create enough money to buy it back again. Take this great power away from them and all great fortunes will disappear, for then this would be a better and happier world to live in. But, if you want to continue to be slaves of the bankers and pay the cost of your own slavery, then let bankers continue to create money and control credit."

Another economic threat comes from industry segments in a country unable to innovate and adapt quickly enough to changes.

Some individuals and companies become complacent and may even refuse to change their practices. We have the American textile industry, the Russian economy, the Japanese pharmaceutical industry, and the German biotechnology industry as examples of this trend. Thus in these segments, the marketplace will displace them with imported products, protective measures notwithstanding.

Resources are very important. Without imported resources, Japan would be a poor, overcrowded nation of fishermen and not much more. We have plenty of natural resources, and they give us strength. However, we are not self-sufficient in all resources, and unless technology provides us with substitutes, we are in need of imports. For example, potash used in fertilizers is imported from Canada.

Day One

This day is already upon us (or possibly even passed us by, depending on how you look at it). The financial crisis that began in 2007 has shown signs of improvement, but does not seem to be ending anytime soon.

The signs of a far more catastrophic economic crisis are staring us in the face. In the U.S. and Canada, we have major problems with older people's retirement benefits and medical care. All that we need is another "market readjustment" like the one in October 1987. If the government tries to save the economy by letting inflation rise, then all bonds, mortgages, insurance companies, loans, and savings would be worthless very quickly. If the government tries to deflate the currency to save the financial institutions, then the economy will collapse. It will take more than bake sales, bottle drives, and garage sales to reduce government indebtedness.

What to do:
- First make sure that you can sit out an economic meltdown. This means that you must have a reserve of food, barter goods, protection, and knowledge and like-minded people around you.

- Convert assets at risk into *hard assets*. Do this in a rational way so as not to lose too much value. Another point, be sure to have small gold and silver coins for smaller purchases. Getting *change* may not be possible in all barter deals.

- Identify people you would like to have on your side when something happens. Cultivate their friendship. Never, ever advertise for survivalists to join you. You may find yourself corresponding with a serial killer, an agent provocateur, or someone who is looking to rip you off.

- Cache some supplies and firearms in case your supplies are taken away.

- Purchase older formularies to make your own medicines, paints, inks, explosives, and other chemical compounds. This way you will have a trade and trade goods to fall back on as the situation worsens.

- Always keep your actual address confidential. Use a company name with a post-office box number or a mail-forwarding service.

- Buy secondhand goods. You will save money and sometimes get really good deals.

Day Two

Most economic meltdown scenarios have two things in common—our greenbacks will be worth a lot less, if anything, and there will be unemployment. Now is the day to decide which of the following scenarios applies and prepare for that scenario. Read these scenarios carefully to assess which of them applies, then go to that section.

Runaway Inflation

Imagine paying two million dollars for a gallon of gasoline while your month's earnings are ten million dollars! Can't happen here,

the government backs the currency, some would say. Our dollar is backed by less than two cents worth of gold. If that does not constitute a lack of backing, I don't know what does.

Let me tell you what it was like in Germany in the early 1920s. There is a famous photograph of a housewife lighting her fire with currency. The real corker was when a woman shopper left a basketful of marks outside a store for a moment. When she returned, the money was still there, but the basket was gone.

With federal debt piling up, who cannot see the obvious consequence? You don't need a newspaper from the future to know that the U.S. dollar is about to take an historic dive and that is just one aspect of the coming financial crisis. The U.S. government has been running up debts in your name for a long time. Those debts will be paid eventually . . . but not the way you think. They will be paid with funny money—paper with green and black ink on it.

The last recession delivered an important warning to Americans—reduce your debts. But it went unheeded. Interest rates fell to the lowest level in decades. Lower rates made it easier to carry debts—especially for the big debtors such as the U.S. government. So, instead of paying off debts, the borrowing continued.

However, interest rates don't fall forever. Eventually they start going up. All of a sudden, the debt level you thought you could afford becomes intolerable, which is why a lot of Americans are downwardly mobile in a big way. They're on the wrong side of the techno-wedge that is dividing the country and the world. This is equally true of governments as well.

Now the stage is set for the dollar to collapse as the government is forced to print extra currency to cover its debts. This won't be the first time that a major government has printed money to reduce its debt, and it won't be the last, but it will be a nightmare for millions of Americans. As the dollar falls, stocks and bonds will continue their downward spiral. The retirement savings of millions of people will be wiped out. Mutual funds will be hardest hit.

Of course, remember that none of this has to happen to you. Here again, there are two separate futures available to you. It's your choice. There are many fine investments that will protect you from the financial crisis that is coming, while earning high rates of interest.

Day One in this scenario can last from one or two days to many years depending upon your economic circumstances. If you are of modest means, it will take time to amass the resources to ride this one out. If you are better off, invest now in hard assets that will keep their value no matter what happens to the currency. For the last 5,000 years, gold and silver kept their value. During the Roman Empire, you could buy a small farm with 100 ounces of gold; you still can do this today. Although gold and silver do not pay interest, they keep their value. Platinum is for the more specialized investor with a larger asset pool. Precious stones are a poor investment. You buy retail and sell at wholesale, 'nuff said.

All countries, China in particular, are improving their living standards. This will cause greater consumption of foods, goods, and services. Raw material stockpiles are low all over the world in part due to multinational companies using the Japanese "just in time" delivery systems. These factors are inflationary in nature. Would the Japanese buy American government bonds if they need more iron ore at higher prices?

In summary, the worst investments to have during runaway inflation are:

- Investments with a fixed return, such as bonds.
- Being a mortgage holder, a lender with a promissory note.

Day One

At the beginning of an inflationary period, more workers are hired, so unemployment is reduced. This initial euphoria results in increased spending. We expect some inflation during boom times. We even welcome it as we can pay bills, mortgages, and loans in cheaper dollars.

What to do:

- Prepare a bug-out kit for each member of your family or group.
- Purchase U.S. (pre-1965) or Canadian (pre-1967) silver coins for *small-change purchases.*
- Purchase bullion gold coins for your major savings.
- Have a three-month supply of the foods you normally eat.
- Have a six-month supply of grain and other bulk foods plus the tools to process them for each member of your group.
- Learn a skill, trade, or profession in demand so that you can barter your services for the items you need.

Day Two

This starts when people realize that inflation is here to stay, and the people living on fixed incomes, i.e., senior citizens, welfare recipients, investors living off bond or bank interest, are placed in a position of abject poverty. This has happened recently in Russia to a lesser degree. Crime will increase as many people pushed into poverty try to survive. To overcome this, more police will be hired and security companies will see a rise in business.

This is the day when companies who have hired workers lay them off, bringing a rise in unemployment. The banks and other lenders will raise interest rates on mortgages and loans to protect the principal. An unprecedented number of foreclosures will follow.

What to do:

- Form a cooperative where you live or move where people with like ideas congregate.
- If many members of your cooperative are unemployed or underemployed, you may want to set up a security company.
- Close out any safety deposit boxes you may have.
- Purchase as much food and other essentials as you can with the currency you have on hand. Store them in a safe place.
- At this point, you should have a guard or guards at all times where you store your goods.

Day Three

Even the bureaucrats will eventually realize that evicting the whole population into the streets will not solve the problem. The politicians will declare a moratorium on all debts and payments and will ration food, gasoline, and other supplies. The black marketeers will hold useless currency, and direct barter will be the fashion of the day.

If the utility companies have to curtail power generation and the delivery of natural gas or fuel supplies, there may be a move to communal housing on the part of the authorities.

What to do:

- The cooperative should form scavenging units to collect bulk food, paper for fuel, and other essentials.
- Group protection must be established so that your home/residence is secure while you are scavenging.
- If you own property, evaluate its security potential. With a moratorium on debts and payments, its value has been dramatically reduced. You may want to move to a safer location.
- When you barter, do not display all barter material you have. For example, if you have sixteen jars of honey, show only one or two. Have the rest out of sight. This has a twofold effect. First, you can say that you are running low on honey and as such get a better trade for it. Second, it reduces the risk of some people wanting to rip you off as being the "honey king."
- Buy non-hybrid seeds. Hybrid plants do not produce seeds. Planting non-hybrids will eliminate the need to buy from seed companies every year.

Day Four

As the situation deteriorates, individuals and groups will take to looting and raiding other people's homes. The government will impose travel and other restrictions on the population. Enforcement of the regulations will be by martial law. A dusk-to-dawn curfew is likely in urban areas.

Stadtrat:

No 0203

Gutsch... der
Stadtgemeinde Traun...

ehn M...

Nur gültig in
die Stadtgemei...
lösung späteste...

No 02178

Gut=Schein
der Stadtgemeinde Traunstein
über
Million Mark
den 3. November 1923

Stadtkämmerei:

Zehn Millionen Mark

Gutschein der Stadtgemeinde Traunstein
über

Zehn Millionen Mark

No 1590

1923.

500 M...

Hauptkassa:

The inflation will result in high prices and shortages for all imported goods, and even used items will have a premium value. Drivers of imported cars will have problems sourcing some parts for these vehicles.

What to do:

- As the old saying goes, "If you are going to Dodge City, know the back way out." You must determine whether to stay put or to leave at this point. No government will have sufficient troops to completely seal off each municipality.
- Make sure that you have a reserve of water, toilet, and waste disposal facilities independent of the public system. It may be a chemical toilet, a full septic system, or as rudimentary as a supply of plastic shopping bags.
- Start to obtain the now-worthless coins and currency. Get as many different kinds and as large quantities as you can. Do the same for military medals and badges. Sooner or later, people will start to collect them, and at that time, it will be another source of funds for you.

Day Five

At this point, the government will more or less give up. Many of the raiding parties will have organized themselves into major gangs. You must have a protective organization around you to counter their numbers. In effect, you must organize *your own government*.

The government will try wage-and-price controls to halt the slide in the dollar's purchasing power. The response to this will be shortages, and black marketeers will emerge.

What to do:

- You should have a *neighborhood protection force*. Do not join one that will not use force to defend its territory. You may be away when your wife and other members of your family are visited by some gang.

- It is time to see whether a new government or a new currency is emerging. If not, it is time to look at long-term food production.
- Plant your garden.

Day Six

A new currency is introduced. It may be regional or national. For example, Nevada could introduce the *Nevada silver peso* or something else that has some value. It may even come to the point that the currency is backed by corn, oil, or whatever is of value or produced in that region.

The currency may be a fiat currency with wages and prices set by the government. This kind of currency works only in a dictatorship having extensive police presence.

What to do:
- Do not exchange gold or silver reserves for any of the new paper currency. In due course, it may go the way of the old one. However, if the coins contain a constant and verifiable silver or gold content, then use them and accumulate them as time passes.
- Keep informed. The official news sources may not tell you the whole truth or even tell you any truth at all. Listen to shortwave radio broadcasts to see how the rest of the world views the changes in America.

Day Seven

Now you will see if the new currency is stable. For example, if the new silver dollars contained one ounce of silver and six months later they are hard to find, that would indicate hoarding or the lack of additional sources of silver. Similarly, if after six months, the dollar contains only a quarter ounce of silver, the new currency is losing its value. This means that you are back to barter and "real money" (gold and silver).

What to do:

- In order not to deplete your hard assets, form a group to exchange services for supplies. If you do not have a skill and don't feel that you have the aptitude for one, become the salesman, delivery man, or whatever is needed in the newly formed group.

- Another way to obtain additional supplies is to bring abandoned farms back into production. If your cooperative is large enough, you may want to send out reconnaissance parties.

Day Eight

Slowly, to some very slowly, a stable new currency will emerge. Be careful that it is not a form of cuckoo-clock money. People traumatized by runaway inflation will support very conservative central bankers. This conservatism will last until a new generation forgets the lessons of the past.

The new fiscal policies will create stable long-term jobs. A pent-up demand for goods and services will create a steady market. People will save money and will avoid debt, so inflationary pressures will be held at bay.

What to do:

- Be watchful always. If you see that paper money is being printed to cover deficits, by now you should know exactly what to do.

- It is time to invest in corporations that are oriented toward the utility sector and providing of services. In a recovery period, they usually do very well.

Day Nine

Be watchful, keep hard reserves on hand, and make sure that the government is not a top-to-down government. Our constitution was designed so that the people are the government. Keep it that way. The problem will be to see whether the constitution survives a runaway inflation. In search of economic stability, we may end up trading off freedoms.

What to do:

- Maintain your emergency stocks, rotate them, and add new items if they are needed.
- See what shortages exist and see if you can procure or make the items in need.

Day Ten

By now you should be out of the economic troubles. However, be vigilant, and keep a skeptical attitude toward government announcements. What you have just come through more or less will be like the German inflation after the First World War.

Stagflation

The expression "stagflation" is composed of two words, stagnation and inflation. Current economic theory states that inflation most likely occurs during an economic upturn, whereas during a stagnation period, inflation is supposed to decline. In a period of stagflation, people are laid off while inflation continues to increase. The resulting scenario will see people on fixed incomes having to eat pet food to meet their protein requirements.

You will see the death of public services. Local, state, and national governments are running out of money. Technically, they are already broke. In the last two years, eighty-five cents out of every dollar borrowed went to the government. It's one thing to borrow heavily when rates are low. However, when rates rise, a heavy debt load is a death sentence.

Government budgets are slashed. Maintenance is cut—and streets will become no more than broken pavement and dirt. Buildings will burn down for lack of fire protection. The urban landscape will not be a pretty sight, and the death of public services will make cities and suburbs even less desirable places to live.

Private companies will step in to do the work that public services used to do, and you can profit richly by investing in them. Neighborhoods will hire their own police forces and waste removal companies. Private education companies will take over the bureaucratic, inefficient public schools. Virtual reality, a new technology, will totally change the way children learn—a monumental opportunity. However, in between, you must muddle through and that is what this section is all about.

As inflation increases, the interest rate on government debts will rise, too. This will require additional funds to cover the interest payments. Bad enough by itself, this is magnified if additional monies are borrowed. The added money required to cover interest payments will be taken from the budgets for maintenance, health, and welfare. So in effect, federal and state governments will download the responsibility to municipal governments, but not the money to pay for new local costs. This downloading will put pressure on the local property tax base, and a patchwork system will emerge depending on the financial resources of each community.

Day One

You will find that every month you have less and less *disposable income*, while at the same time nothing much seems to change. When you start hearing of police and fire stations closing, laying off of public service employees while the government is still borrowing heavily, then Day One is near.

The prices will change slowly, and the inflation will be reflected by higher interest rates, thus adding to the government's balance of payments problems. This will reduce government expenditure for

services as a higher percentage of the revenues are used to service the debt.

What to do:

- Carefully examine the type of community you live in. Look at the percentage of people living on fixed incomes. Examine the makeup of your neighborhood and decide whether to move or stay.
- Start to accumulate barter goods.

Day Two

At this point, you will see policemen and policewomen moonlighting as security guards when off shift. As a matter of fact, there will be such a boom in the security business that you might consider becoming a member of that profession. Strikes will most likely be staged for job security rather than for increased wages.

As interest rates for mortgages and loans increase, the number of foreclosures will also increase. The empty factories with no activities will not pay taxes, adding to the pressure on government revenues.

What to do:

- As suggested earlier, you should form a cooperative for protection and security.
- The formation of a security company will enable you to earn money and at the same time train your own security force.

Day Three

The slow, steady decline continues. You are worse off each day. The laid-off government workers will add to the welfare rolls just when the money received buys less every day. Desperate people will turn to crime, and theft will be rampant. Lack of money will slow the justice system, and the accused will continue their criminal activities while on bail.

People reduced to poverty will join extremist parties that offer fast and quick solutions, usually wrong. The rise of these extremist parties will be like what happened in Germany and Italy in the 1920s.

What to do:

- Now that you have a cooperative, start organizing parties to collect food, fuel, and other supplies.
- If there are other cooperatives in your area, arrange to barter supplies and services. This cross trade will be even more important in the days to come.

Day Four

One of two things will occur, runaway inflation (you are back to the previous scenario) or further decline. About this time, pet stores are closing for now people will eat the pet food if not the pets. The problem most governments face is that in order to pay the interest on the national debt, they must keep inflation down and as such the interest rates. Most of them do this by letting the unemployment rates rise dramatically.

The changes will accelerate, thus further adding to the misery of the middle and lower classes.

What to do:

- Your cooperative should prepare for continued decline and have the ability to grow your own food.
- Establish a rotation of jobs to keep your area clean and habitable.

Day Five

About this time, the government will declare the emergency to be almost over. Do not be fooled. They have just admitted that they have no idea how to deal with the situation. It will continue to slide. The slide will be gradual. Many of the highly specialized unemployed workers will turn to providing basic services like cutting firewood. The stagnation will continue, although at a slower pace.

A cash-strapped government will cut back on foreign aid and social programs. This will add to the misery worldwide. The tide of illegal immigration will increase, the U.S. being the favorite target of immigrants in search of a better standard of living.

What to do:

- If you have adequate protection, you may want to admit people with tradable skills to your cooperative. This is called a skills-for-barter system.
- The additional skills can be bartered for other needed essentials and supplies. You should be able to pick up power tools at fire-sale prices, and if you have electricity, you are now an industrialist.
- You must be self-sufficient in food supplies. Potato soup may be bland, but it is nutritious.
- If you have babies, make sure that you have baby food on hand or can make baby food. Many adult foods are too harsh for a young child's stomach.

Day Six

Given the stagnation and the erosion of the tax base, the government may declare that everyone must perform a number of hours of public service per week. Try to organize so that your members perform law enforcement, medical, and other skilled trades as your public-service responsibility.

High prices for imported goods, including crude oil, will result in reduction or elimination of government services requiring the use of vehicles. These services will include the police and fire departments, garbage pickup, rural mail deliveries, snow clearing, road repairs, municipal transport, and other services we take for granted today.

What to do:

- Try to have one member of your group in the job assignment office.
- If you can have the work in your area assigned to the members of your cooperative, this will enhance the security for the members.

Day Seven

The government will be trying to create real jobs in the public service sector. This will mean that fire, police, and other public services will gradually improve. However, the trend will be to use a manual rather than mechanized approach, thus adding to employment. This will also mean that many people who were previously unemployable because of lack of skills will now have jobs.

What to do:

- It may be possible to start a private "labor exchange." This way you can find jobs for the members of your cooperative and at the same time earn additional funds as a "finder's fee" or alternately contract your services out.
- Form a corporation to supply specialized services. This is needed because an hour's wage is drastically different between a doctor and a laborer.

Day Eight

A new type of society will emerge. The thrust will be toward full employment while at the same time minimizing government expenditures. This will mean excellent municipal services in some parts of a region and none in others, all depending on the local tax base and the ability of people to pay. The divergence of the quality of local services will result in a call for some kind of federally mandated minimum level of services. Given the rough ride government went through, we are not likely to see too much federal intervention.

What to do:

- In a full employment situation, a premium is placed on skilled labor. If you have not already done so, upgrade the skills of those in your cooperative.

- Do not dissolve your cooperative. Such a cooperative can help you reduce expenditures by bulk purchases in the best of times, and you can maintain a neighborhood protection program.

Day Nine

The most likely outcome of this scenario is a government with very conservative monetary policies and very little in the form of "social safety nets." This will give rise to a two-tiered society. There will be the haves, with access to medical care and money for their old age. The have-nots will have to do without food stamps and welfare benefits so crime will rise and so will the police and security functions.

It is likely that democracy as we know it will change with rights being based on merit. In other words, if you do not contribute to society, you will not be able to vote.

Currency Collapse

Should all our foreign creditors demand repayment of the monies lent to us in their currency, our dollar would collapse overnight. You may say, "So what? We just pay them in devalued dollars." Look at the downside of this scenario. Recent reports indicate that 10,000 affluent Americans a month are leaving the U.S. in response to tax hikes and disturbing expansion of federal powers. Even some multi-billionaires have forsaken U.S. citizenship to become citizens of another country, like Kenneth Dart who now is in the Caribbean tax haven of Belize.

The flight of the monied class means that those remaining will have to pay even higher taxes to make up for the taxes lost to the government. There comes a point of diminished returns, when the cost of collecting the tax will equal that of the tax collected.

Higher taxes lead to an underground economy where people work for cash, and in turn, this reduces the taxes available for the government. Italy provides a vivid example of this type of economy and its resilience. The Italian tax police are active, but they make only a small dent in the underground economy. Many companies are incorporated in Liechtenstein for the same reason—tax avoidance. Most governments realize that large multinational companies can evade excessive taxation so they shift the tax burden onto the people working for wages. The attitude toward the working stiff was summed up by an astute man as, "Once you have them by the balls, their hearts and minds will follow." What does this have to do with currency collapse?

In order to control people, most governments encourage the use of direct payment cards, credit cards, and bank accounts, all for a simple reason—the ease of tracking where your money comes from and where it goes. The trend is toward elimination of money as we know it today. This desire, coupled with the availability of smart cards, powerful computers, and the cooperation of financial institutions, will lead to an ever-increasing government scrutiny of your lifestyle. People react in different ways. Some form barter groups where you earn barter hours of services or goods. Some turn to gold and silver, and some will not use banks or credit cards. The net effect is that the government, increasingly desirous of controlling private transactions, will institute some kind of cashless society.

Then we have an ever-increasing government debt, much of it financed by foreigners, mostly by China and Japan. Pension funds and insurance companies are also holding large chunks of our debt. Should the stock market boom come to a halt, many program traders will have to dump government securities to shore up their holdings.

A catastrophic crop failure could also bring on a currency collapse. Lack of food exports would undermine the nation's ability to bring back home our greenbacks spent on cars and computers.

Day One

When you hear slogans saying "Buy American" in a big way and the consumer receives special treatment such as sales tax rebates for purchasing U.S.-made goods, then you know that Day One is near. When interest payments to foreign creditors exceed 20 percent of the government's total expenditure, then we are coming even closer to the brink.

Another sign of a currency collapse will be if oil, gold, and other commodities currently quoted in U.S. dollars are quoted in Euros, British pounds, or Japanese yen. This will signal that we are on the edge of the abyss. In some ways, recent administrations looked upon the devalued U.S. dollar as a way to boost exports and to repay foreign loans with devalued money. This is only partially right. As long as we import more than we export, the dollar will continue to weaken, and we will pay more for imported goods. This will be especially detrimental when American-made items are unavailable. For example, almost all computer chips are made by a handful of companies in Asia. We must pay the higher price if we are to make computers.

What to do:

- Start a savings program first by accumulating U.S. and Canadian silver coins, then by purchasing gold bullion coins. The silver coins will be used as "small change," and the gold for major purchases.
- Remove currency, precious metals, and similar items from your safety deposit box.
- If you can afford it, buy some travelers checks denominated in Swiss francs. These do not have an expiry date. Or you can actually buy currency. Currencies of the following countries are

generally deemed relatively safe: Switzerland, Japan, Sweden, Canada, and Australia.

- Update your immunizations.

Day Two

Barter will start in a big way. You will learn the value of the three Gs—guns, gold, and grub. In addition to the three Gs, other items will have the value of currency. If you have a surplus, you can use them as currency. Do not try to sell goods at inflated prices. That will create resentment, and someone may do you in. Trade or barter items. In case you want to stock up on them, experience has given a list of *currency* items. The priorities and desirability change slightly from place to place. These are:

- Liquor
- Tobacco
- Drugs (including antibiotics)
- Ammunition (mainly .22 LR)
- Matches
- Fishhooks
- Lighters and flints
- Razor blades
- Firearms
- Feminine hygiene supplies
- Birth control pills

We are dependent on imported high-technology goods from other countries. Whole industries may be brought to halt. A glaring example is the computer industry. If chip imports are curtailed, we will realize our dependency on international trade. The country most affected by a U.S. currency collapse would be Japan, not because of their U.S. currency holdings, but because their national economy is so dependent on trade with the United States.

What to do:

- Stock up on the barter items listed above.

- Invest in domestic producers of antimony, arsenic, asbestos, barite, bauxite, bismuth, cadmium, cesium, chromium, cobalt, columbium (niobium), diamonds, fluorspar, graphite, indium, manganese, nickel, platinum group metals (the good stuff in your catalytic converter), potash, rubidium, oil, strontium, tantalum, thallium, tin, tungsten, vanadium, yttrium, and zinc. Talk about being dependent on foreign trade!

- If you can, go into business for yourself. This way you can take advantage of changing conditions.

Day Three

By this time, expect the government to be saying that America is still strong and we do not need imports as we can make anything. Do not believe this for a moment. For example, you will find that items made out of stainless steel will become very pricey. The U.S. is almost entirely dependent on imported nickel and chrome in the production of stainless steel. Although there are some reserves in the U.S.—for example, there is a ferronickel smelter in Riddle, Oregon—the mining and refining process is much more expensive than using the imported metal or ore.

At this point, you will know that gold is people's money and paper is banker's money. That is little comfort to the little guy who has been wiped out. It is probably too late to convert your dollars into gold and silver. Food prices have risen dramatically.

Isolationism will first reflect itself in the form of protectionism, a very popular solution. This will appeal to the people as they will see it as a sign of independence and resistance to oppression of foreign traders.

What to do:

- You can form a group to scavenge for metals in short supply. We already recover gold from computers and electronic devices. You may want to collect used machine tools, scrap aluminum—go after anything containing tin. Many ointment tubes are made out of tin.

105

- This is possibly the last time you can purchase gold legally. If you have more money, do so now.

Day Four

Given the large quantity of U.S. currency distributed worldwide—about 70 percent of the U.S. currency is held outside the country—the attack on the American currency will have international implications. It may even hurt those who started the assault in the first place. We may face an attack by several powers to redeem U.S. currency in their hands with gold. Countries with very large holdings of U.S. currency are: Russia, China, South American countries, the Middle East, Southeast Asia, and the former Warsaw pact countries. A very impressive list of potential enemies!

What the government may do is issue a parallel currency. This would be of different colors for domestic and foreign uses. Any American dollars would have to be exchanged when the U.S. border is crossed. The U.S. dollars acquired overseas would have to be converted into domestic dollars, and of course, such earnings would be taxed.

What to do:
- Cache your excess supplies. These caches can be in the old-time cold cellars of farms you occupy, in the fields, or any other place where the goods are safe and protected.
- Time to add to your supply of seeds.
- To conserve your assets, you should be a supplier of some goods or services in demand. Perhaps the best way to go about this is by forming a cooperative or a group of like-minded individuals.

Day Five

Our federal government promises to back the currency with gold and silver. To do this, the government will declare the private holding of gold and silver a criminal offense and will require citizens to turn these in for paper dollars. There will be an amnesty period

of several months to do this. This will please foreigners holding U.S. greenbacks.

U.S. investments abroad may be sold off to redeem U.S. dollars held by foreigners. There may be pressure from foreigners to peg the U.S. dollar to gold. Since the U.S. is a large gold producer, this may appeal to many foreign governments. What many fail to understand is that 100 years of U.S. gold production would be required to make a major dent in our debt to foreigners. Talk about long-range programs!

What to do:

- Cache your precious metal assets.
- Add more food to your cached stores.
- You can add to your precious metals cache by recovering them from electronic equipment.

Day Six

The precious metal grab will not work. People are not that stupid. There will be some show trials of people found with precious metals in their possession. This may even include a widow or two wearing their wedding rings! The government will start a search of coin dealers' records to identify those who have bought gold coins. They may use credit card records to identify those purchased precious metals in the past.

What to do:

- Since you do not want to spend your time in some public institution making license plates, it is well to be wary of people wanting to make trades for gold or silver. Be wary of strangers.
- Formalize some kind of barter system with other groups in your area.

Day Seven

This has much in common with runaway inflation in the sense that money is virtually worthless. The foreign holders of U.S. currency will force their governments to seize the remaining American assets in their countries. The prices of imported goods will continue to rise.

What to do:
- As suggested under runaway inflation, get as many of the worthless coins and currency you can for future trade and collector purposes.
- You can make money by scavenging parts from foreign-made cars and appliances and repairing them. If you have the space, you could even make some part-time money scavenging appliance parts and repair manuals.

Day Eight

The U.S. may mobilize to recover assets. If this happens, we are in a war situation and the scenario becomes one of warfare survival. The government may issue a fiat currency to replace the ailing U.S. dollar. The prices may be set by the government in the currency, and trading on the black market may become a criminal felony.

In a warfare scenario, we would have very few allies. The countries we could count on would be Canada, the United Kingdom, and perhaps Israel.

What to do:
- During a major war, there is a curtailment of civilian availability of certain foods, goods, metals, and other items needed to carry out the war. Get into obtaining these items and supplying them to the private sector. Never do this in a way that you are diverting supplies earmarked for the war effort. There are many small producers and manufacturers who do not come under the war procurement boards. Deal with them.
- Scavenging for imported metals will help the country while providing extra income for you.

Day Nine

If war is declared, go to the appropriate scenario. But if American mobilization is sufficient to cool the governing heads in other countries, you have another worry. This is the question of what the government will do to stabilize the currency. We may see a two-tiered system where the foreign dollar holdings will be backed by something more than the goodwill of the U.S. government. Usually in such cases, there is a cutoff date by which these dollar holdings must be presented to foreign central banks for eventual redemption.

Another possibility is the establishment of a cashless society. Instead of money, machine-readable cards would be used, and as such, totalitarian control over the citizen is achieved.

What to do:

- Try to benefit from this situation. If you have U.S. dollar holdings in foreign banks, see if they are covered by the new two-tier system.
- Time to practice barter. Trading does not leave a paper trail, and your acquisitions may be safer than if you bought them by using currency.

A Real Depression

According to the economists, the signs of a coming depression are:

♦ A decrease in bank credit

- A decrease in debits in banking demand deposits
- A decline in public debt
- A decline in the velocity of money
- A decline in crude materials prices
- A decline in mortgage debt

THE SURVIVALIST'S HANDBOOK

At present, we do not have many of the above happening. However, the Great Depression came with frightening speed. Our politicians and their economists tell us that we cannot have a replay of 1929. They say that with social security, Medicare, welfare, and the Federal Reserve System in place, a depression can't happen. Whenever a politician says that we can't have a depression today like in the thirties, when he says that the likelihood is like a man having a baby, then I say watch out for pregnant males. When the savings-and-loan collapse resulted in a $600 billion bailout by the federal government, many said, "See, it can't happen here, no run on the banks." Guess whose grandchildren will have to pay off that bailout?

How can we have a depression? A depression is a mild form of economic collapse. We have all seen pictures of people taking in each other's laundry to create economic activity, bankers selling apples on street corners, and soup lines. Once again, you can avoid these. With a little foresight, you can even profit from a depression. Who profited from the last depression? The people who bailed out of the overheated stock market, the ones with *liquid assets*. They were able to buy houses at fire-sale prices. They could hire people to work for them at low wages, for they were solvent.

There are analysts who point to the deflationary effect coming from the cycle analyses, in particular the Kondratieff wave. Kondratieff waves are long waves that were first experienced between the 1780s and 1810s. This was when revolutionary advances were made in spinning and weaving of cotton, when canals and modern roads were built, and when the steam engine was first introduced as a source of power. The second long wave came between the 1840s and 1870s, being produced by the worldwide construction and the development of revolutionary steelmaking processes that drastically reduced costs, so that steel became a feasible construction material.

The third great wave began in the early 1900s and carried forward till the late 1920s. It was generated by the widespread intro-duction of electrical power, by the development of the automobile,

and by the emergence of the modern chemical industries. The fourth wave started in the 1960s, and we do not quite know when it ended in the 1990s. We saw the exploration of space and widespread introduction of computers. Following each of these waves, there occurred a period of slacker business activity when the economy *digested* the technological advances introduced during the expansionary period. We are in a slack period now.

The wave theory suggests that the best investments will be in government bonds, treasury bills, and cash. This is true in a depression. But a depression may start off with an inflationary spurt before the crash. If you lost one-half of the value of your bonds due to inflation, then you will have suffered already. It is better to have gold and silver on hand.

When does a recession turn into a depression? That is an academic argument, and what we have to focus on is how to pull through whatever economic ills beset us. Some are saying that the recession of 1992 was actually a major readjustment of the economic system, like the start of the Industrial Revolution. If this is the case, then manufacturing jobs are going to give way to a search for the perfect hamburger flipper. Although a service economy absorbs the untrained, it also results in a general reduction of earning power and as a consequence in our standard of living.

Day One

If you hear that the government has actually reduced the national debt by 1 percent, then Day One may or may not be here. The start of a depression is as hard to pin down as the start of runaway inflation. Later, with hindsight, we will point to all the signs, but then hindsight is always 20/20.

Much to the delight of bankers and lending institutions, government debt is never retired, thus interest payments continue on the accumulating debt. These interest payments may be as high as 80 percent of a government's expenditure in Third World countries to as low as 30 percent in the more industrialized countries.

What to do:

- Sell off stocks selectively after checking the price/earnings ratios first. Those which have a p/e ratio in excess of ten should go first and so on. However, there are certain stocks that are almost depression proof. Do your research, and hang onto those stocks.
- Have cash reserves on hand.

Day Two

Nobody is buying capital goods. Real estate offices have whole streets up for sale. Manufacturing companies are laying off people. Stores are closing. Banks are foreclosing. These are the signs that tell you almost surely that a depression has begun. Crime will not increase significantly during this period as money buys more and those on fixed incomes, pensions, and welfare will be better off.

Mutual funds will be the first to experience the problems. Since so much pension money is tied up in them, the future for those relying on pensions will be bleak. This will undermine the pension income of many people.

What to do:

- Have most of your assets in cash.
- Add to your hard-metal reserves, mostly in the form of silver coins. This is a hedge against the situation turning into runaway inflation.

Day Three

Stockbrokers are not jumping out of the windows of their high-rise headquarters, partially because the windows cannot be opened and mostly because they have no personal commitment to their clients. The first to be hit by falling stock prices will be the private investors. These people usually get in the market late and as such do not have the large initial profits to draw upon to ride out the stock market depression.

To reduce government expenditures, the government will lower welfare payments, negatively index pensions, and thus those on fixed incomes will see a downward shift in their spending power. If it is an election year, the present government is likely to be voted out of office. The new one may introduce work schemes, but not follow up on them.

What to do:

- Certain professions and trades are always in demand. Learn a skill like carpentry, electrical work, or metalworking.
- This is about the right time to form a cooperative to ride out the depression.

Day Four

The government will try to put people to work through make-work schemes. More money will have to be printed to do this, and this will create an inflationary spiral. The employment created will not greatly reduce the number of unemployed. But there will be inflationary pressures from the new money printed. There may be forced relocation of the unemployed into areas deemed to have a shortage of labor.

People unemployed and those on reduced pensions will demand that the government do something. As the corporate profits disappear and personal income tax revenues tumble, the government will see its ability to pump money into the economy evaporate. The government will start a panic downsizing. The interest on government debt will remain the same so it will take a larger percentage of revenues to cover this fixed expense.

What to do:

- When you hear that this is happening, convert your cash into hard assets and liquidate your bonds.
- Avoid relocation. Keep employed even if you have to hire yourself. Forced relocation will result in many people becoming virtual serfs.

- Form a corporation to provide you with a vehicle for investments and job creation.

Day Five

Being afraid of a runaway inflation, the Federal Reserve will increase the discount rate or even require banks to have a larger percentage of their deposits lent to the Federal Reserve System. However, the Federal Reserve is owned by banks, so do not expect too many moves on the Fed's part to hurt its owners. This will further depress the stock market. The remaining mutual funds will go under. The mutual fund collapse will erode retirement benefits for most people. Currently many pension plans are heavily invested in mutual funds and mortgages. Those holding mortgages will see their earnings disappear as people out of work will not be able to make their payments.

The depression should bottom out at this point unless people decide to change the system completely. In that case, you probably will see a civil war or anarchy emerging.

What to do:
- Start to invest in depressed real estate and other real assets. It may be time to purchase stocks.
- You may want to increase your holdings of gold and silver as a hedge against a currency collapse. This is a possible event during this scenario.

Day Six

There will be a short period of stabilization in stock prices. Many people with fixed incomes and retirement benefits will feel better off once again. However, this is dependent on the solvency of the institution their funds are invested in. We will see some minor price rises in some commodities, but production will remain in a slump.

Depending on whether the rest of the world is in a depression, this period could see a major increase in U.S. exports, thereby reducing our foreign debt. The increase in exports may force other countries to lower the value of their currency in terms of the U.S. dollar, thus reducing the export advantage. This may lead to the introduction of tariff increases on selected groups. High tariffs will deepen the depression.

What to do:

- It is time to form some kind of contracting business. This way you can be flexible about using other people's talents as you don't have to carry them on your payroll.
- You may be able to contract your services out to governments to do the services the governments are downsizing.

Day Seven

Most make-work schemes will be stopped to reduce government borrowing. The private sector will see an increased demand for its goods, and people thrown off government programs will be largely absorbed in the private sector. However, this is a slow process, and in the transition phase, many will be worse off. The fact that people are being hired will give hope to many and reduce the pressures on the fabric of society.

What to do:

- Your contracting business could bid on work that had been done by the government workers in-house. Downsizing in the public sector will still require that certain public services continue, only now these will be contracted out.
- Maintain contacts with the remaining government employees to know what opportunities are available for you. Remember, one man's downsizing could be another's opportunity to provide a well-paying service.

Day Eight

The government will see its deficits reduced, and the dollar or whatever passes for money will strengthen. Reduced government borrowing will make more money available to the private sector. It is at this point that you will see the first improvement in the lifestyles of the people. The changes will be tentative at first, but they should gain momentum as more people are employed.

The pent-up demand for goods and services creates a very vibrant economy. The challenge will be to control inflationary trends.

What to do:
- With an active contracting business, you can cash in on the initial business expansion in a big way. It is also a good way to put members of your cooperative in a money-earning capacity.
- You should have contacts in the regulatory agencies to keep you ahead of changes in regulations.

Day Nine

An economic upturn is well on its way. The government coffers are getting fuller by the day, mostly due to more people paying income taxes. This will add to reduction in the accumulated deficits. The government may even stop borrowing entirely. The additional employment will reduce government expenditures for welfare and other social security benefits, freeing up money for capital projects.

What to do:
- Get involved in the political process to ensure that we don't slip into an overheated economy by printing money and enlarging the government.

Energy Wars

The fundamental problem facing the North American governments is a transition from cheap imported oil to a more balanced system of energy sources. It is clear that current domestic resources—oil, gas, coal, and nuclear—cannot be greatly increased in the near future.

If you don't think that crude oil is important to us, look at the table on the next page! We will be short of more than gasoline if oil imports are cut off!

Different petroleum end products require different ships to carry them across the high seas. That's expensive. Instead, oil companies made the decision to transport crude oil and refine it at home ports, close to consumer markets. This meant the arrival of the age of the supertanker and super ports.

Throughout history, when man has had an abundance of any commodity, he has tended to waste it because he views it cheaply. U.S. postwar society was predicated on inexpensive oil. Our electrical generating plants were built to its specifications. So, too, our cars, homes, and many other aspects of our daily lives were constructed with little regard to fuel economy. A series of other factors operated as well.

After the Santa Barbara oil blowout, we temporarily cut off continental-shelf exploration. We found enormous reserves on Alaska's North Slope, then delayed construction of an 800-mile trans-Alaskan pipeline because of largely justified environmental concerns.

In 1954, the Supreme Court held that under the 1938 Natural Gas Act, the Federal Power Commission had to regulate not only what interstate pipelines could charge for the gas they sold to local utilities, but also how much the pipelines could pay when they bought gas from producers (the wellhead price). In the years following, this latter tool was used to regulate interstate prices at artificially low levels. As a result of this underpricing, the best fuel we have from an environmental standpoint is woefully overused—and misused as well, in such areas as industrial power generation.

Finally, our development of alternate energy sources moved forward at a snail's pace because, for one thing, extracting gas or

UNLEADED

4 1 5

LEADED
LUS

4 2 9

CME

4 3 9

elf se

oil from coal (syngas and syncrude) is enormously expensive. And, for another, we assumed that nuclear energy would be the ultimate answer, so we did not look elsewhere. Wind and solar were deemed to be special scenarios and attracted very little research money.

The alternate energy sources for vehicles are electricity, ethanol, methanol, natural gas, and synthetic fuel. Converting to these is expensive, but the biggest stumbling block will be the energy distribution system. It is one thing to put new tanks in the municipal garage for the buses. It is a whole new venture to equip service stations with them. Another problem is that the replacement fuels are less efficient than gasoline or diesel oil. This requires more frequent refueling of the vehicles or larger fuel tanks on them.

A partial solution was the introduction of ethanol/gasoline blends. The downside is that only 10 percent of the gasoline is replaced by ethanol. To add more would require major reengineering of the vehicles. A 100 percent ethanol fuel would necessitate that we plant corn even on the strips dividing the interstate highways.

In this scenario where you live is very important. If you live close to an oil field, refinery, coal field, or hydroelectric facility, your options may be wholly different. Let us assume that there is adequate electric power generated in your area. In that case, opting to have an electric vehicle, as they become available, is the correct choice. Therefore, get to know what energy supplies are available in your area as a first choice. If none is available, remember that during the Second World War, methane gas generated from chicken manure provided fuel for many a civilian car. There is always an energy source available if you look hard enough.

The problem with raising money for changing our energy mix is that heavy government borrowing *crowds out* the corporate borrowers from the marketplace. Thus, changing the mix to favor more domestically produced energy is endangered by the deficits successive administrations have run up.

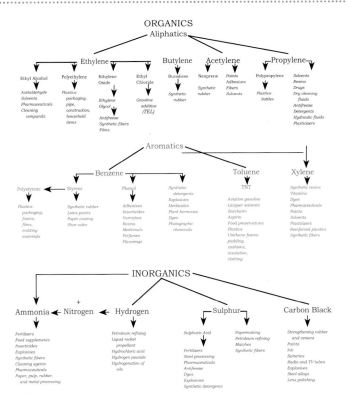

Source: Facts About Oil, American Petroleum Institute
THE PETROCHEMICAL FAMILY TREE

Day One

Within recent memory we have an example of a mild form of Energy War. Remember 1973 and 1974? That was only a mild money grab by the OPEC countries. We really did not have an oil shortage. Compared to what is likely to happen, those days were only minor skirmishes. What to do in a major energy confrontation? You won't

123

be roadkill as traffic will be very sparse. Fuel rationing will mark the arrival of Day Two. However, it is more difficult to define the signs of Day One.

After the 1974 oil scare, we have had conservation measures, a search for alternate fuels, crash programs into developing solar, wind, biomass, and even tidal energies. Today, more than twenty years later, what have we? The only sign of energy conservation use are the smaller cars with better fuel efficiency. We have wasted the intervening years. What if Saudi Arabia goes the way of Iran? We are even more dependent on imported oil, and at the same time, the major oil reserves are in unstable, if not hostile, lands.

What to do:
- Prepare a regional alternate-fuels study.
- Obtain publications pertaining to alternate fuels.
- Have the necessary tools and equipment to convert your vehicle to an alternate fuel.
- Have a reserve supply of fuel on hand to carry you through the transition phase. Be sure that the fuel reserve does not present a direct fire hazard to your living quarters or your stored supplies.
- Form some kind of informal understanding with a local fuel distributor. It would be nice if you could countertrade for your services or some supplies well in advance of Day Two.

Day Two

Now that you have a definite shortage, try to assess whether this is a blip or a long-term shortage—a very difficult task to do. If you believe that the shortages are a definite trend, start switching over to your chosen alternate energy source. The governments will react as before. For example, odd-numbered last digit on the license plates fill up on Mondays and so on, but no major grass-roots changes are expected. We will hear about reducing travel by car, carpools, and the like while our politicians jet around the world in search of fulfillment.

You will not be very wrong to assume that the days of cheap petroleum are ending and the competition between the chemical industry and the fuel needs will inexorably push prices up.

What to do:
- If you live in a cold climate, have small space heaters on hand.
- Purchase a wood-burning stove, and install it.
- Invest in railroad stocks.

Day Three

The government realizes that it has dwindling stocks of energy on hand. Drastic conservation measures are introduced. If the energy shortage is in the form of gasoline, the most likely event, ration books will be printed.

Railroads will enlarge their systems and may even switch over to coal. The nation's truckers will face ruin. It is likely that only short-distance trucking companies will survive. The railroads' profits will ensure that they can successfully compete with all other forms of transportation systems. Airlines will further raise their prices, and more people will take to using the railroads for their transportation needs.

What to do:
- Have some fuel supplies stored in a safe place. Do not store fuel in your residence. First there is the fire hazard, and second, if a container is punctured, the fumes may be toxic.
- With a little cooperation of friends, you can probably pick up some additional ration coupons or turn to the black market.
- Consider moving to the country.
- Add alternate energy sources to your supplies. These can range from solar panels and windmills to methane generators for your car.

Day Four

Black markets will flourish, and the government will react by issuing special stickers allowing cars to be used only for necessary trips.

These tags or stickers will be twofold—permanent for essential workers and temporary for special cases. Use of vehicles may be controlled by roadside checks or even by roadblocks. This will reduce consumption by 25 percent.

The authorities will try to limit travel by people. This may come by permits for essential travel only. The increased use of coal will lead to increased pollution. Those who can will try to move out from smog-ridden cities. This will result in a reduced tax base for the city's coffers, and as a result, municipal services will be reduced.

There will be an accelerated development of all energy resources. These will bring on other problems. To give you an over-view, let us look at these energy sources one by one.

Oil—The national petroleum reserves will be developed. Increased offshore activity may result in increased pollution.

Natural Gas—Gas production from difficult-to-reach formations. Additional pipelines will be built.

Coal—Larger mines and relaxation of air-quality regulations.

Nuclear—Streamlined siting and licensing procedures will reduce reliability.

Synthetic fuels—These require huge quantities of water, thus will add to the water shortages.

Shale oil—Will require modifications to Colorado's environmental lands and will divert precious water resources.

Geothermal—Very limited areas are suitable for its development.

Solar—The twenty years lost due to our complacency may not be all bad. Instead of massive collectors that take up acre upon acre of desert space or ugly water heaters that clutter up so many rooftops today, we may have a technology emerging. Increasing the efficiency of solar energy conversion while simultaneously lowering the power requirements of manufacturing, solar may become a sufficient power source of energy for both home and industrial use. The newly emerging molecular nanotechnology may make this a reality. Thus oil may become as obsolete as whale oil was in the early twentieth century. Chief anticipated feedstock materials for molecular manufacturing are carbon, nitrogen, oxygen, and hydrogen, in other words even grass can be used.

What to do:

- Obtain all the stickers you can. Initially the criteria for issuing stickers will be rather lenient. That is the time to get them.
- Have a library of conversion plans for alternate fuels for your vehicle.

Day Five

If the shortages continue, commercial, institutional, construction, and manufacturing use of oil will be reduced. This will reduce consumption by another twenty-five percent. We import more than fifty percent of our fuel needs, so this and the previous measures will prove to be inadequate.

This is the time when a switch over to coal and coal-derived fuels is likely. Coal mining is more labor intensive than crude oil production, and coal miners earn more money than laborers. To compound this, in underground mines only sixty percent of the coal is recovered in contrast to 90 percent in open-pit mines. To offset this increased yield, open-pit mines bring along with them a host of environmental problems. These range from creating eyesores through an increase in flash floods. The high sulfur content of many coals requires expensive scrubbers in coal-burning power plants.

Oil exploration in the U.S. diminished due to the low return on investment and a relatively high risk. We will see an increased exploration activity all across the nation to find additional reserves. More gas fields will be brought into production, requiring more pipe and construction equipment. In turn, more energy will be needed to manufacture the new equipment.

What to do:

- Form an energy research company with other like-minded people, and obtain government contracts.

- Shop cooperatively. That is pool-purchase to obtain better prices and at the same time make only one trip to get your shopping done.

Day Six

At this point, the government must assess whether domestic supplies are sufficient to fuel the nation's remaining vehicles.

They probably will not be sufficient, so a crash program to extract oil from coal will be restarted, just like in Germany during World War II. This will work, but the resulting fuel will be expensive, and there will be an environmental downside in the form of water and air pollution associated with production.

We may see coal gasification plants being set up in areas not yet blessed by natural gas pipelines. Pollution control measures will be further relaxed. As a matter of fact, we may see a total decline in environmental measures.

What to do:

- Get a job in the energy sector.
- Arrange a barter system for fuel supplies.
- There are many local coal, lignite, and peat outcrops. These can be exploited with minimum capital investment. Small deposits are labor intensive, but labor is likely to be plentiful under the circumstances.
- You may want to consider converting vehicles to run on methane generated from chicken manure. This was common practice during World War II in occupied countries.

Day Seven

Agriculture will suffer. The energy input is very heavy in that sector in the form of diesel fuel, gasoline, natural gas (where available), energy-derived fertilizers, and chemicals. The government will provide special energy supplies to the agri-food industry. The result will be an increase in food prices, and we may even see some local shortages.

Another area of special fuel allocations will be in the military sector. A black market in fuel will flourish, and this will lead to further government regulations. A new bureaucracy will be established to control and monitor fuel allocation plans. This, of course, will be nothing but another opportunity for black marketeers to put more people on their payrolls.

What to do:
- Be a farmer or have farmer associates. Keep in mind that a typical family without tractors and other mechanized equipment can farm only about ten to fifteen acres. If you were part of a food cooperative, then you probably will need all members working on the farm just to keep it going.
- Obtain books dealing with alternate energy devices from earlier days. These will help you prepare alternate sources of energy.

Day Eight

If the energy crisis is not solved by this time, you should be ready for an alternate lifestyle. Bicycles and carts will form the mode of transportation. We may even see the return of the horse-drawn wagon. This will add to the number of people employed, while at the same time the cost of food and goods will increase to reflect the higher cost of transportation.

The lowering standard of living will see a nostalgia craze for the bountiful 1950s. Home entertainment will rise as people will travel only on special occasions.

What to do:
- You may want to have a supply of tools and parts for bicycles.
- Learn about bicycle repair.

- Reduction in travel will give rise to increased neighborhood shopping. The corner store will make a comeback.

Day Nine

It may be that now only government vehicles have a steady fuel supply. As a result, road maintenance will be minimal and what is done will be limited to major roads and the interstate system. The worsening of transportation networks will result in shortages of foods and supplies required by the cities. Small towns and villages may be completely cut off from the distribution system.

We will rediscover the use of waterways and canals for transporting goods. This is by far the cheapest way to move people and goods. The use of water transport will see new life breathed into port cities. Employment will rise as the use of machines decline. This reduction in the use of machines will result in a reduced standard of living for many. The gap between the haves and have-nots will increase dangerously, thus raising the specter of conflict between classes.

What to do:
- If you can organize unemployed transportation workers, you can start a small cartage outfit to deliver goods to the rail-heads.
- Consider raising horses.

Day Ten

This is not the end, only a bend in the road. The country has turned a corner, and we are relying on domestic energy resources. This will strengthen the dollar, and America will be well on its way to retaining its superpower status.

Climatic survival scenarios range from the very cold ice age to the very hot ozone layer depletion. We already practice a mild form of climatic survival as the seasons change. In the winter, we add more clothing and turn up the heat. Yet at the same time, the climate can kill us. For example, if you've watched enough movies, you're familiar with the scene where the hero takes off his shirt to walk through the desert under the midday sun. I really can't think of an easier way of committing suicide. Climatic survival revolves around maintaining body temperature, water content, and protection from harmful events. Having defined climatic survival, the next step is to decide how to go about achieving it.

We already have clothing for seasonal extremes. You should have a larger than average supply on hand and, most importantly, sturdy footwear. While you can scavenge most other pieces of clothing, footwear is more size specific and as such harder to obtain. Break in your footwear before you really need it for a trek in the wilds.

Most of the climatic scenarios are interrelated. For example, when the El Niño current changed off the west coast of South America, we had storms and accompanying floods on our West Coast, fewer hurricanes on the East Coast, and heat waves in the Midwest. Scientists believe changes in El Niño are caused by windshifts in the upper atmosphere. Similarly, volcanoes can deplete the ozone layer and lead to global cooling. The thin layer of atmosphere protects us from the chill of outer space. We are doing our level best to diminish it.

The scenarios laid out can be caused by nature or man. Mankind now can cause major changes in the climate, as we have demonstrated in the past century. Whether this comes from pollution or diversion of rivers is immaterial. We are impacting on the planet's ecosystem, and in many cases, we are unaware of the changes we have set in motion.

Day One

Today is the day. You should have equipment, supplies, and knowledge to ride out climatic changes. You already do this in a seasonal way, so continue. Make sure that your clothing keeps its insulating value when wet. Wool is best for this. Also have clothing suitable for hard use. When choosing clothing, keep in mind that you may have to sleep in it for days on end.

There will be a number of scenarios dealing with climate, even though some are in other sections of this book. All of them require that you have a food supply on hand *before* the event. This can't be emphasized enough. A solid food supply and means of purifying water are musts for all scenarios.

What to do:

- To repeat, always have a small emergency kit with you. You may get tired of reading the same advice again and again, however, it is too important to ignore.

- Have good, sturdy footwear suitable to the climate in your area.

- Invest in clothing suitable for the expected climatic maximums in your area.

- In the wintertime, have an emergency kit in your vehicle—at least a space blanket, candles, beef jerky, extra socks, a pullover, boots, and whatever medications you need to take.

- Carry in your car a topographical map of the area you are traveling and have a city map of your destination.

- Be aware of the elevations along your route. Know your roads, and be prepared for climatic events.

Day Two

This is when you will see which of the following scenarios apply. Go to that scenario.

Severe Ozone Layer Depletion

The ozone layer protects us from the harmful effects of the ultraviolet radiation emanating from the sun. Should the ozone layer completely disappear, life on earth would be extinct. For example, ultraviolet lamps are used in hospitals to sterilize certain areas; ultraviolet kills germs, and without germs (bacteria), life as we know it would be impossible. One of the worst scenarios we can contemplate is a complete disruption of the ozone layer. Should the ozone layer disappear, even fish in the ocean would be affected.

We already hear about *holes* in the ozone layer over the Arctic and the Antarctic. These holes are huge. To date, their size has been of interest to scientists only, mainly because they are located over largely unpopulated areas, but also because those living in those lands already have to wear heavy clothing as a protection against the cold and sunglasses due to the chance of snow blindness. However, there were some reports from southern Chile citing the presence of blind sheep and other animals. Imagine this on a larger scale in more populated areas of the globe. The UV (ultraviolet) Index is part of most weather reports. This is a recent phenomenon.

How is the ozone layer depleted? Scientists have determined that chlorofluorocarbons (CFCs) and nitrous oxides destroy ozone by converting it into ordinary oxygen. CFCs are found in air conditioners, aerosol cans, and cleaning solvents. Nitrous oxides are released through chemical fertilizers and the burning of fossil fuels. (CFCs were supposed to be eliminated under the Montreal Protocol, which established a phased-in halt in production of CFCs.)

Another new threat to the ozone layer is methyl bromide. It is used by farmers, who inject this gas into the soil, then cover it with plastic. Methyl bromide kills insects, weeds, fungus—you name it. The EPA lists it as a Class I acute toxin—the most deadly of all substances. It is not only toxic, but as the gas rises, it breaks down into bromide. Bromide transforms ozone into oxygen, further attacking the ozone layer. The United Nations Environment Program meeting in Montreal in September 1997 called for a total ban on the use of methyl bromide.

Volcanic activity also leads to the reduction of the ozone layer. Sulfur dioxide thrown into the atmosphere during a volcanic eruption enters the stratosphere, and a small amount of it is converted into microscopic particles of sulfates. Many of these sulfates are carried to higher altitudes where they combine with chlorate products from CFCs and cause an even-more rapid ozone depletion.

It is estimated that for each 1 percent reduction in the ozone layer, the incidence of skin cancer rises by 2 percent, and the relationship is not linear. That is to say that at a certain point, the skin cancer rates would be endemic worldwide. At this time, we do not have precise information concerning the reduction in crop yields, but a reduction will occur.

Day One

When you hear about a 50 percent reduction in the ozone layer close to where you live, you know that Day One has begun. Reduction in the ozone layer results in reduced crop yields. The skin cancer rates will rise over time. Since the ozone layer depletion is over sparsely populated areas, the public input will be muted.

There will be the usual denials about faulty observations, a blip in the statistics, and the like. All this posturing will achieve is to put off the day of reckoning.

What to do:

- Reduce your family's exposure to the sun.
- Have on hand plenty of sunscreen. Even if you do not use it, it can be traded later.
- Have really good sunglasses to avoid cataracts in later years.
- Have plenty of the food you eat on hand.
- Wear light-colored clothing, wear it loose, but be sure it is tightly woven.
- Wear a hat with a wide brim. Be careful of reflected ultraviolet rays from water and other reflecting materials.

Day Two

Prolonged exposure to increasing amounts of UV radiation will result in an increased number of skin cancers and will affect *all animals* and *all plants* to some degree. Blind sheep are not able to look after themselves! There will be a black market in CFCs by developing countries. Initially this will be overlooked by the developed countries, but as the situation worsens, pressure will be applied to stop the production and trade in CFCs.

The Montreal Protocol aimed at eliminating the use and production of CFCs will have teeth added to it.

This elimination of CFCs will result in price increases from $0.50 per pound to more than seven dollars per pound, putting a higher price on refrigeration equipment essential to preventing food spoilage. The poorer countries will go right on manufacturing CFCs.

What to do:

- Learn all you can about UV protection.
- Get plenty of zinc oxide to make sunscreens. If you don't have any, you can always extract it from pennies (1983 and later years).
- Learn all about food preservation.

Day Three

If the ozone layer depletion continues, there will be a drastic move to so-called safe areas. Whole populations may be on the move. To avoid this, study microclimates in the affected areas. This research may pay off handsomely if you can move to some part of your region that is subject to frequent cloudy days. Seattle and Vancouver, Canada, are examples of such a microclimate.

The authorities' desire to eliminate the smuggling of CFCs will cause them to give police agencies and customs additional powers at the expense of personal freedoms.

What to do:
- Obtain meteorological data for the region where you live.
- The decision to move or not to move should be tempered by careful review of your lifestyle. If you are a farmer, you may want to keep animals inside barns and have larger crop areas to make up for the yield reduction.
- Obtain geological maps of your area. Find out where caves and abandoned mines are. This information may be a lifesaver in later days.

Day Four

The government will take frantic measures to get the ozone up where it is needed, in the stratosphere (nine to thirty-one miles above us). What we call the weather is in the troposphere, which extends only nine miles above the earth's surface.

Increased levels of UV-B radiation reduces the ability of the body's immune system to fight foreign substances that enter through the skin. For example, infection of *herpes simplex* may weaken the body's ability to fight subsequent infections. This will add to the health care system's overload.

The poorer countries will not be able to afford the newly developed substitutes for CFCs so food will spoil and thus even further reduce foodstuffs, in particular, the proteins available to the population.

What to do:
- Go into the ozone-rebuilding chemical business.

- Add to your food supplies, since the price of food is sure to go up. The increases in food prices will not be even. First to go up will be for foodstuffs sensitive to UV rays. UV-B radiation damages fish larvae and juveniles as well as the phytoplankton base of the food chain.
- Another good business will be making closely woven fabric for use in areas with high UV radiation. Hat making will also see a resurgence.

Day Five

The ozone layer may stabilize or continue degrading. Given our current efforts, the ozone layer is likely to degrade. Our governments are reactive rather than proactive, meaning we react after the event has happened rather than making efforts to avoid it. By this time, people will have a very good idea of what to expect. All government pronouncements will be treated with healthy skepticism. The escalation in food prices will continue.

What to do:
- By this time you should have an "earth-sheltered" house. That is a house partially underground, much like the basement of an ordinary house. This reduces energy consumption while shielding you from the ultraviolet rays.
- You may want to change your working hours, perhaps even working at night to reduce exposure to the UV rays.

Day Six

If the ozone layer depletion continues, you will have to take extreme measures. These may include relocating. Remember, the areas where the government will relocate you will be crowded and you will end up in a refugee camp, with all that entails. A police-state apparatus will be associated with all government camps. FEMA (the Federal Emergency Management Agency) will take control of local police forces.

Shortages will force the government to ration food. Prices will continue to rise, and a black market will flourish. Beef herds will probably be slaughtered to free up grains for feeding people.

What to do:
- Think about getting into indoor greenhouse or even hydroponic gardening. You can protect your animals by keeping them in a barn with PVC roof skylights.
- Before you relocate, think about security while en route and what you will lose in the move. If the situation is extreme, you may have no choice but to move. During the move, make sure that your lead vehicle is capable of ramming through roadblocks.

Day Seven

This is the time when you can see large-scale results of the ozone layer depletion. These will range from very poor harvests to the collapse of fishing in shallow waters. The problem with this phase is that the overpopulation in many places, particularly in urban areas, will increase food shortages made worse by poor distribution systems. The reduction in food supplies will bring along other major problems, like those discussed under the heading of "Food Wars."

Overpopulation, coupled with poor medical infrastructures, will lead to a very large death toll. A combination of this and reduced food supplies will cause a reduction of population to a point where some kind of self-sufficiency is achieved.

What to do:
- Review your options in terms of moving your living quarters and activities underground.
- Fish farming under UV-resistant cover may be a good business now.
- If you live in an area with caves, make use of them.

Day Eight

If the ozone layer keeps diminishing, the population will have to move underground and hydroponic gardening will be the only viable long-term means of growing food. Along with people, domestic animals will have to move underground. This move will create untold misery in poor countries. The death rate in some of the affected countries will skyrocket. Based upon current knowledge, countries most at risk will be Peru, Bolivia, Chile, South Africa, Namibia, Indonesia, and the Philippines.

The reduction of the ozone layer will have other impacts on the population. Certain foodstuffs will virtually disappear, especially those grown at higher elevations.

What to do:
- In an underground home, your heating and cooling requirements will be reduced. The resultant energy savings may be used to start a small cottage industry or a maintenance facility.
- Have a moat or a ditch around your property to keep out blind animals.

Day Nine

If the ozone layer disappears, life on earth's surface will be impossible. It will be like living on a hostile planet. The only remaining life will be in the deeps of the oceans, where the sun never shines. This is a certain doomsday scenario.

The Greenhouse Effect

We often hear about increased carbon dioxide emissions leading to the *greenhouse effect*. Let us take a look at what is at stake here and what we can do to protect our well-being.

Global warming can affect all our lives. The atmosphere is a blanket of air that surrounds the planet. It traps the heat from the sun's rays on the Earth's surface and controls the temperature. This is the natural greenhouse effect, which keeps the planet at optimum temperature for all living things. Greenhouse gases in the atmosphere, such as water vapor, carbon dioxide, methane, chlorofluorocarbons, and nitrous oxide, trap the sun's heat and warm the earth's surface. Without this effect, the average temperatures would be $-1°F$ ($-15°C$), instead of $59°F$ ($15°C$).

Carbon dioxide amounts are increased in the atmosphere by animal respirations, burning of fossil fuels (such as gas in our cars), volcanic activity, and deforestation. Methane is released when vegetation is burned, digested, or left to rot (for example, grazing cattle and rotting materials in landfills).

Global warming is the result of too many greenhouse gases in the atmosphere. They trap the heat being reflected from the earth and cause the temperatures to rise excessively. As a result, some of the areas of the continent will receive more rain and others will become drier. Coastlines may flood, and islands may go under water because of the melting of polar icecaps and glaciers. Thus, plants and animals will have to adjust to this new climate. Many of them may become extinct.

There are two schools of thought on this subject. One believes the above scenario. The other believes that increased greenhouse effect will increase the earth's albedo (reflectivity of the sunlight we receive) and as a result we will slip into another ice age. This will lead to greater swings in the weather: more storms, hurricanes, and typhoons.

In both cases, some areas will get drier and some will get wetter. The prairies and the Midwest are expected to get drier and, as a result, food productivity will drop. The sea levels will rise as the Arctic and Antarctic ice melts. This will put many of the existing harbors at risk. All port cities with elevations of less than fifty feet will be at short-term risk. The trees we see thriving in the south are going to move north, and some species will die off as a result of this northward migration. Another plant migration will occur in alpine areas, where some species will be crowded out and will have to move higher, resulting in some alpine species being crowded off these areas.

The wild variations in the weather patterns will produce more storms and blizzards and make life more difficult. We know very little about the ultimate outcome. The so-called fossil record points toward the coming ice age scenario. The coming of an ice age is not a slow, gradual process. The increasing snow covering will increase the reflection of the sunlight back into space, and the following year will see an even-greater spread of the snowfields. After a couple of years of this, the southward exodus of the population from the northern areas will be a massive movement.

Increased carbon dioxide in the atmosphere can stimulate plant growth by increasing the rate of photosynthesis, although shortages of water and other nutrients may limit yields. Some crop plants respond more to carbon dioxide enhancement than others, usually depending on how they photosynthesize. Wheat, barley, and potatoes respond quite vigorously to carbon dioxide enhancement, but maize, sorghum, millet, and sugar cane do not. Also on the negative side, the food quality of some plants will decline, and insect damage will increase.

Another problem with the increased levels of carbon dioxide is found in the oceans and lakes since cold water absorbs more carbon dioxide than warm water. The net result of this is the lower layers in still waters act as a huge repository of carbon dioxide. If these layers are warmed up or disturbed, carbon dioxide is released to the atmosphere. Recently in Cameroon, Lake Nyos had a shift in the lake bottom. The released carbon dioxide wiped out several villages, the people dying because carbon dioxide displaced the air. What happens if a warmer atmosphere melts ice sheets, causing seismic activity to increase and stir up the colder waters, is a frightening thought.

Although governments are talking about reducing carbon dioxide emissions, the powerful lobby of energy companies and oil producers stalls any meaningful action. As a result, we have Day One nearing. According to some, it is already here.

Day One

The problem with the greenhouse effect is the same as with an ice age—you don't know when it starts. By the time you know that it has started, you are well into it. An international conference held in 1996 officially confirmed that global temperatures are up. So we are in Day One now.

A vivid example is found in a recently released study pertaining to the MacKenzie Valley in the Northwest Territories of Canada. There, the average temperature increase of 3.2°F (1.8°C) resulted in melting of the permafrost, drying of lakes, and increased incidence and severity of forest fires. This is happening in an area deemed to be least likely affected by the greenhouse effect.

What to do:

- Keep current on what the scientists are saying. Read *Scientific American, Nature, Science,* and other popular scientific publications. If you can't afford to subscribe, read them at your local public library.
- Talk to knowledgeable people about changes in the local weather patterns, and find out if the changes are new to the area.
- If you hear that there were major crop failures in your area for two or more years in a row, this is a certain sign of weather pattern changes.
- Have a reserve of food as described under other scenarios. Remember, in any disaster scenario, if you have food on hand, you will not have to expose yourself or your family to degenerates and improvident individuals who will be doing each other in. You will be able to wait and see how the situation develops.
- Read up on substitutes for imported foods, and learn to make them.
- Take several camping trips with your family, and live off emergency food sources. In this way, you will find out which of the emergency foods you like and how to prepare them under various circumstances.

Day Two

This will begin with raging scientific debates among the specialists. In the meantime, the government will say that what has happened is only a blip. Keep your eyes open. If major government construction is going on in the mountains of Georgia (which was ice free in the last ice age), you can bet that the government is preparing for a coming ice age. Weather patterns will be even more extreme. According to one major scientific theory, the Midwest and the Canadian prairies may have reduced food yields, thereby greatly reducing food surpluses.

What to do:
- Obtain a power generator. It's best to have one that runs on diesel fuel since you can always use heating fuel in it.
- Add to your food reserves by forming a cooperative of like-minded individuals.

Day Three

Now that we are well into the greenhouse effect, the governments will belatedly take steps to curtail the use of fossil fuels. At first it will be more talk than action, but as the situation worsens, they will really start to curtail energy use. In the beginning, we will see restrictions on vehicle use, followed by a "carbon tax" on utilities and oil companies. The carbon tax will surcharge electricity generated from fossil fuels, gasoline, diesel fuel, and natural gas in proportion to the carbon dioxide released when they are burned. This will lead to at least a doubling of energy costs. The rising energy costs and lower crop yield will cause many farmers to give up their way of life and become dispossessed people.

International cooperation will begin to reduce carbon dioxide emissions. We may even see some kind of international army created to enforce these reduction measures.

What to do:

- Increase your food supply in stages. First obtain imported staples like coffee, then other foodstuffs.
- To reduce your energy costs, use solar heating and electric power generation, if you can.

Day Four

Increases in the carbon tax will continue, putting miners, refinery workers, transport employees, and others out of work. Many coal-fired power-generating stations will either close or curtail electricity production. Brownouts will be normal.

The melting of the polar ice caps could cause sea levels to rise by 200 to 400 feet. New York, Los Angeles, London, Rome, Amsterdam, Antwerp, Marseilles, Helsinki, Stockholm, Brussels, Bombay, Karachi, and scores of other cities would be underwater. Cattle farms in Siberia and on the Canadian tundra would be common.

What to do:

- Carbon taxes may also carry with them carbon credits. This provides a business opportunity. For example, if you are a corn farmer, your crop absorbs carbon dioxide. This absorption, less the carbon dioxide generated working the field, will be trade-able to those paying the carbon tax.
- If you have the contacts, you may be able to set up an agency to trade carbon credits.

Day Five

At this point, food production will be reduced due to marginal farmers going out of business. The government will create special carbon credits for farmers. These credits will carry with them tax rebates for the fuel used in farming.

The oceans will not be able to cope with the increasing carbon dioxide load. When this happens, oysters, clams, and other shellfish will be decimated. Coral reefs, including the Australian barrier reefs, will crumble.

The Northern Hemisphere will experience more rain and higher temperatures. Weeds like heavy rains. They will flourish, putting up to 40 percent of the crops at risk. Heavy rains also encourage the growth of fungi and other agents that prey on our food supply.

What to do:
- As said before, associate with a farm or farm group.
- The need for rodent and moisture control will increase. Food storage will become very important. Be ahead of the coming need by going into this business now.

Day Six

At this time, you will see whether we are entering a new ice age or a tropical planet phase. Either will bring major changes with it. In an ice age scenario, the southern U.S. will be habitable, but the northern half will have an ice sheath covering it. The ice age may arrive in as short a period of time as three years. In a tropical scenario, many crops will migrate northward, squeezing out the cold-weather trees and plants. The spread of these plants is slow. You will have more than 100 years to adjust to this warm cycle.

What to do:
- Establish contact with scientific groups, and be on the distribution list of position papers prepared by those groups.
- Subscribe to scientific journals. If you can't afford to subscribe, visit your public library at regular intervals.

Day Seven

If an ice age is coming, the government will institute a plan to relocate people. Initially, these will be voluntary, but the later stages will be compulsory. If you gotta go, go early. The relocation will cause massive abandonment of production facilities in the threatened areas. Shortages of all types will be the norm.

What to do:

- Have contacts at decision-making levels. See what the local authorities are doing with their families. If they are moving them, so should you.
- Within your cooperative, agree in advance where to meet if you have to move.

Day Eight

The relocation of people will generate "refugee camps," where people will be housed until a new home is found for them. These camps will be run by the authorities and may degenerate into labor camps as the situation worsens. They will also serve as recruiting bases for troops and police forces of the government. The people recruited will have no pity for the population outside the camps.

Life will become brutish for many. People on the move will strip the countryside of food and supplies. Those living in the areas stripped of food will have no choice but to join the migrants.

What to do:

- Avoid the "refugee camps" like the plague. Have a group around you, and buy or stake out a property that is suitable for food production. Keep in mind that you may have to defend your new home.
- By this time, you should have a small group of like-minded individuals around you. Make use of their skills and talents to start a new society.

Day Nine

This is when the new society emerges. Most of the people are completely dependent on government handouts, and their lives are run by the government. The only free people will be in the fringe areas, and even these will be squeezed out as the ice sheaths grow. However, keep in mind that the government's control will be very amorphous. That is, the further you are away from the bases of control, the less control the authorities will have over you.

Nuclear Winter

A widespread nuclear war with a number of surface bursts will throw an incredible amount of crap up into the atmosphere. We have evidence of what happens when dust levels in the atmosphere increase. For example, major volcanic activities have resulted in some very cool summers. The Krakatoa explosion in Indonesia affected North American climates for years afterward. With the breakup of the Soviet Union, the chances for a Total War have receded. However, we must wait to see whether China will become a trading nation or an imperial power with territorial ambitions.

The unsettled Sino-Russian border region could pose a threat of a major war. The Russian war machine is in such a sorry state that it would have to resort to the use of nuclear weapons very early in such a conflict. The problem America is facing is that both powers would throw a few ICBMs with nuclear warheads our way just to make sure that there would be no superpower left at the end of that catfight.

Don't believe it? For a minute place yourself in the shoes of the Russian and Chinese leaders. In a nuclear exchange, there are no winners, just losers of different degrees. Most wars are fought for territorial acquisition, and the loser almost always becomes an occupied land. After a major nuclear exchange, what is there to occupy? Blighted lands with demoralized survivors suffering various degrees of disorientation. Not a very profitable scenario for the winner, especially a winner that will be severely disabled as well.

An intact United States will ensure nothing but American dominance over the world for a long, long time to come. There is no other power capable of projecting power like the U.S., be it in the form of aircraft carriers, marines, prepositioned supplies, or whatever is required to project that power.

To illustrate this point, look back upon the results of the two World Wars. Britain was bled white, and today it is just another squabbling member of the European Community. Its empire is long gone. Only the shared history and system of laws provide some sense of continuity. The average person in the United Kingdom is looking to move to the United States, Canada, or Australia. And those two wars were fought with conventional weapons. Just think what a nuclear confrontation would bring.

In a nuclear winter scenario, we are plagued with fallout, cold weather, and loss of harvests, just to mention a few of the nasty things happening at the same time your world is falling apart around you. The old Mormon ethic of having a year's supply of food on hand will become a necessity. Scavenging will become very difficult, so all that you need should be on hand before the nuclear winter comes. Exposing yourself to radiation to obtain supplies will shorten your life span considerably. People without radiation-measuring instruments will be at a great disadvantage, and those with them will be able to trade information for supplies.

Day One

If you observe that the Russians and the Chinese have many armed border conflicts, then you should prepare for the possibility of this scenario. Long before this, you should have obtained access to a fallout shelter or prepared one for your family. A root cellar with a three-foot earth cover will protect you from most radiation. Your shelter should be airtight and have a high-efficiency air filter to keep out fallout particles.

Another warning sign will be nuclear conflicts between India and Pakistan, Israel and Syria, or an attack launched by either Iran or Iraq on Saudi Arabia. These smaller-scale confrontations would at best produce a *Nuclear Autumn*. However, this is enough to cause widespread starvation.

What to do:
- Assemble a library on nuclear warfare survival.
- Read up on fallout shelter design and construction.
- Identify potential fallout shelters where you work and where you travel.

- Assemble a bicycle-operated power generator. This can be used to recharge batteries for your radio, radiation detection instruments, and other things.
- Lay in supplies of potassium iodide and L-cysteine. Look for details in Chapter Five.

Day Two

They did it. America may or may not be a nuclear-blighted land. If there is radiation in your area, go immediately to your fallout shelter. If you don't have a shelter, find a shielded structure, and wait until the radiation dies down. Depending on your distance from Ground Zero, you may have several hours to seek shelter or improvise one.

Shielding can be increased by adding mass to your shelter. This mass may come from earth, sandbags, books, and newspapers. Glass has virtually no shielding value, save keeping out radioactive particles, providing that it is intact. Keep in mind that three feet of earth will provide almost 100 percent protection from radiation.

What to do:
- Keep a detailed log of fallout readings. Find out if the fallout is from weapons or from a blown-up nuclear reactor.
- If you are in a hot spot or faced with fallout from a nuclear reactor, you must move.

Day Three

With all that fallout in the atmosphere, it will be very difficult to plan for the future. If you find that the reduction in radiation does not follow the *Rule of Seven*, then you are either in a *hot spot* or are receiving fallout from a reactor. The Rule of Seven is simple: After 7 hours, the radiation should be one tenth, and after 7×7 hours, it should be one hundredth, and so on. If that is not the case, you must relocate. This is the time when you will have evidence of the start of a nuclear winter.

Survivors must now be in shelters. Those without shelters will have absorbed lethal doses of radiation if they were in a target area.

What to do:

- If you evacuate, take a radiation reading every ten minutes. If you have a dosimeter, read it every hour. Keep a log of these readings.
- If you are in a shelter and don't have a remote probe for your radiation meter, you may have to take a reading outside. Make your excursion as short as possible. Take several readings at different areas around your shelter. Keep a log of all readings.
- Decontaminate any items brought into your shelter.

Day Four

Now that you have had a year without summer, it is time to take stock. A likely outcome will be a witch hunt for scientists and others deemed to be responsible for making nuclear weapons possible. Offer refuge to scientists, which will allow you to corner the technological market sure to emerge eventually.

Many undernourished people will die because they have no insulation from the cold. A diet deficient in fats will add to this death toll. For example, a diet relying on rabbits results in "rabbit starvation." The more you eat, the less nourishment you get, making you less and less resistant to the elements.

What to do:

- Keep listening to your shortwave radio, and plot on a map the worldwide effects of nuclear winter.
- Plot the decay rates and see if they follow the Rule of Seven. Again, the rule is that is after seven hours, the radiation should be one tenth, and after 7×7 hours, it should be one hundredth of the initial dose.
- Keep a detailed log of weather observations. This will help you see which pattern evolves.

Day Five

You have emerged from the first winter. The question is whether there will be a second nuclear winter. If the signs point to a continuing nuclear winter, then there will be a mass migration south. Your best bet is to be either first or last to move. In between the casualties will be high.

What to do:
- Plant a small test plot with a selection of your seeds. The results will show you what may thrive in the new climate.
- Observe the condition of the plants and animals around you. This will enable you to see what damage the fallout has done.

Day Six

Lousy crop yields and a further loss of livestock will mark this phase. Many of the remaining survivors, facing starvation, will join raiding outfits to obtain food.

During the early raids on farms, many farmers will lose their lives. The newcomers will not be able to farm successfully, further reducing the available food supplies.

What to do:
- Plant a larger garden and keep a record of what pests are eating your crops.
- Form a cooperative to defend your supplies.

Day Seven

The weather should improve. Some kind of government group will emerge to map the radiation levels in the country. Some regions will start to organize a form of government. We will see an increase in bartering among these regions.

By now we should have a good idea which animals in the wild have survived. It will be too early to see if mutations are emerging.

What to do:
- Participate in the radiation survey by making available your radiation logbook to others.

- Keep a detailed log of weather-related data, such as wind, temperature, and snowfall.
- One of the major concern among the survivors will be locating family members. You can assist in this by placing bulletin boards at swap meets.

Day Eight

A new civilization emerges and worry about ozone layer depletion, the crap in the atmosphere, and the long-term genetic impact of the radiation received by the survivors will be addressed. In the past days, these concerns will have been an intellectual luxury. Now they will be of vital importance.

What to do:

- If you are a scientist, you may want to keep a low profile. The scientists will be blamed for what has taken place.
- The seeds you have been saving will be worth their weight in gold. Treat them accordingly.

Day Nine

International trade and cooperation begins once again. Initially it will consist of ships trading in scavenged goods. In America, it is impossible to predict the type of society that will emerge, but most likely it will be one of two types. The first type may be a dictatorship, required initially to maintain some semblance of civilization. The second type will be characterized by local governments cooperating on mutual issues, but being very suspicious of any type of *big government*. Municipalities are not known for warfare and fabrication of nuclear weapons.

What to do:

- Make sure that your group is represented on the municipal government. You may even want a leadership position.
- In the absence of government, form some kind of a trading alliance with neighboring survival groups.

Day Ten

The new world systems are creaking along. More than likely, any scientific activity leading to weapons manufacture will be banned.

Earthquakes

Sometimes preservation of health and wealth is more important than profit. When a *real earthquake* hits America, the financial and health implications will be extraordinary. For a good example, we have to go no farther than California. Scientists used to believe that an earthquake relieves stresses along a fault line, so people believed that they were safer after an earthquake. Nothing could be further from the truth. In fact, since the January 1994 quake in Los Angeles, the area has been continually rocked by aftershocks, some of which have reached major earthquake strength themselves.

California doesn't just happen to be a place that gets a lot of earthquakes. It does happen to be the home of the San Andreas Fault, a 650-mile long crack in the earth's crust that stretches from north of San Francisco southward almost all the way to the Mexican border. Scientists now know that the San Andreas Fault is the junction of two of the dozen or so interlocking tectonic plates that make up the earth's outer surface. These plates are in motion—they actually float over the earth's hot and semifluid mantle. Moreover, scientists know that the other fault lines that honeycomb California are interrelated. Earthquakes along one fault line are believed to trigger others on connecting fault lines. Although there is a little truth to the idea that an earthquake relieves pressure, in reality California has gone through a 300-year history of relative quiet and the incredible pressure built up must soon be relieved.

Here are the bleak facts:

A study developed by the California Department of Conservation for emergency response personnel predicts that a large

159

earthquake along the Hayward Fault near San Francisco would kill 1,500 to 4,500 people and injure more than 50,000. But that's just the beginning. An inevitable social upheaval would follow and consume more than five million residents of the area, as the lifelines of power, fuel, food, water, sewage, and communications were severed. The social order we take for granted would come completely unglued, and the results would be horrifying.

The prognosis for Los Angeles after a catastrophic earthquake on the San Andreas is even worse. A 1980 federal report places the potential damage to buildings and contents alone at $25 billion (1980 dollars). There would be more than 50,000 homeless, 3,000 to 12,500 killed, and 12,000 to 50,000 needing hospitalization that wouldn't be available. It would be the worst natural disaster in U.S. history. But that is *still not the worst*.

The Newport-Inglewood Fault zone runs in from the sea and between the airport and downtown Los Angeles. An earthquake measuring 7.5 on the Richter scale along that fault would be far, far worse. Building damage could hit $62 billion. The number of people rendered homeless could reach 200,000, the injured 20,000 to 80,000, and the death toll could range from 4,500 to 21,000. And that is just from the initial shock. The nightmare aftermath would be almost unimaginable, and all this in spite of L.A.'s spread-out style.

The flurry of recent earthquakes in California has residents scurrying to stores for emergency supplies. Items such as thermal blankets, crowbars, fire extinguishers, and water purification tablets are moving briskly. The biggest seller, though, is emergency gas-shutoff wrenches. Many stores now carry seismic gas shutoff valves which automatically shut off the flow of gas when a significant seismic event occurs.

The duration of most earthquakes is measured in seconds. The 1906 quake in San Francisco measured forty seconds, and the 1933 Long Beach quake was seven seconds.

California is not alone in being earthquake prone. The Saint Lawrence Fault in northern New York, as well as the history of Missouri and the Midwest earthquakes all point to widespread earthquake activity throughout America.

Although it's still impossible to predict exactly when and where earthquakes will occur, scientists are making great strides with the detection of gases and patterns of microquakes that precede large tremors. So, before too long, we may be able to predict the *big one*. The question is, if officials could predict a big quake, would they tell us? They believe the social disruption caused by the prediction would be worse than the quake itself, especially if they were wrong. Thus *prepare at home*.

What we do know is that when the big one comes, the consequences are going to be beyond anything we can imagine. Besides the direct devastation, the financial aftershocks will ruin people. Insurance companies will default. Businesses will be gone overnight, never to rebuild. Those that are not destroyed will be getting out as quickly as they can find a state to move to and a truck to haul their goods.

The steady stream of people already leaving the state will become an exodus. All the neighboring states will strain to absorb the sudden increases in population. Public services in these states will be strained to the limit as homeless and penniless victims of the earthquake arrive looking for subsistence. Property values in the neighboring states will skyrocket as the law of supply and demand pushes available property prices through the roof. I can only say that if you are anywhere near geographical areas that will be affected, you should plan your life accordingly.

Tsunamis are usually caused by earthquakes in the ocean floor. These deadly tidal waves cause some of the worst natural disasters. They can travel up to 600 miles per hour. Midocean height of the waves is usually less than four feet, but as a tsunami enters shallow water, the wave height may be as much as 100 feet. If you are in the path of one, move to high ground quickly. Tsunamis are a series of waves, so stay out of the area until the "all clear" signal is given by the authorities.

Recently it was revealed that the U.S.S.R. undertook research and even testing of "tectonic" weapons by using underground nuclear explosives to trigger earthquakes. The projects were code-named Merkur and Vulkan. The work was continued under the Russian Yeltsin regime. So Mother Nature may be outdone by man once again when it comes to destruction.

Day One

If you live in an earthquake-prone area, your Day One is already on hand. If you live in a "quiet" area, you may have a little more time. The Richter scale is used to measure the magnitude (size) of an earthquake. In a nutshell, at magnitude 3 you start to feel the tremors, at magnitude 5 some damage occurs, and anything over 6.5 is dangerous.

Even if you are in a so-called safe area, earthquakes can affect you. Many goods and services originate from earthquake-prone areas. So earthquakes may create shortages in areas not directly affected.

What to do:

- Have a small emergency kit prepared and close to you at all times.
- You should not have any heavy objects over your bed. Keep heavy objects lower than the head height of your children.
- If you have time, empty out high bookcases, china cabinets, and other shelves. Remove glass from overhead lights.
- Keep a supply of ready-to-eat foods and cans of fruit juices on hand.
- Your car should be road ready with the gas tank at least half full.
- Many of our computer components originate in California. Examine components in your equipment to determine which ones should have spare parts laid in now.
- Make plans to reunite the family.
- All family members should know how to turn off gas, water, and electricity. Obtain tools to turn these off, and locate them close to the valves or switches.

Day Two

The earth is shaking and quaking. Most injuries and damage come about from falling objects and debris, such as chimneys, plaster, and light fixtures. A community is usually left without power, gas, water, and telephone communications. Fallen power lines and broken gas mains can cause even more problems.

You will be on your own for a while. This is when you will reap the benefits of laying in supplies before the event.

What to do indoors:
- Stay inside. Do not run outside as you may be hit by falling debris. Wear shoes.
- Take cover under a heavy desk, table, or bed, or stand under an inside doorway away from windows. A doorframe is usually a building's strongest point and least likely to collapse.
- Lower floors are safer than upper floors. The upper stories of buildings could come crashing down with you with them.
- Do not use elevators in case of a power outage.
- Open cupboards carefully. The contents may tumble out.
- Clean up dangerous spills.

What to do outdoors:
- Wear shoes. Do not enter any buildings. They could collapse on you.
- Keep away from tall buildings, trees, and telephone and electrical lines.
- Try to make it to the tops of hills. The hillsides may slide.
- Stay away from beaches. A tsunami may be on the way.
- If you are in your car, stop as quickly as possible. Stay in your car. It can protect you from falling debris.

Day Three

After an earthquake, it is time to take stock. Keep off the streets to give authorities a chance to clean up and make emergency repairs. You probably will experience aftershocks, which may be of large magnitude. This period should be used to assess damage and to obtain information on the magnitude of the earthquake.

The extent of damage to your home will determine whether you should pitch a tent outside your house or not. In either case, it is best to stay close to the house to provide some protection to its contents.

What to do:

- Check for injuries, and provide first aid.
- Listen to your battery-operated or car radio to local stations for official information.
- Check for fires.
- Do not enter damaged buildings as walls may still collapse.
- If you suspect damage, turn off the gas, power, and water.
- Do not use the telephone except in a real emergency. Keep the lines open for emergency use.
- If the water is off, you can use water from water heaters, toilet tanks (except those with additives), melted ice cubes, or canned vegetables.
- Check sewage lines before flushing toilets.
- Check chimneys for damage, and note any cracks in them. Do this from a distance, if you can.
- If the power is off, use food from your freezer before it spoils. (A full freezer will keep foods frozen for forty-eight hours.)
- Do not go sightseeing. Keep the roads clear for emergency vehicles.
- Wear shoes when you are walking through debris.
- Be prepared for aftershocks.
- Wear heavy gloves when moving damaged objects.

Day Four

It's time to call the insurance adjuster and listen to broadcasts for information about federal aid and other assistance available to victims. Depending on the damage, you may receive assistance from the Red Cross, as well as federal, state, and municipal agencies. The priority of these agencies will be to free trapped people and give emergency medical aid. Only after that's done will some kind of food distribution scheme be set up.

What to do:
- If you have a video or still camera with film, record the extent of the damage around your property.
- Prepare a list of the damage, and record the serial numbers of damaged equipment.
- Stay close to home. If looters are seen in your neighborhood, try to call the law enforcement authorities.

Day Five

You will start to repair and rebuild.

What to do:
- If you are rebuilding a house, try to make it earthquake resistant.

Storms and Blizzards

The tornado belt is moving north. The first evidence of this was in the sixties when a twister ran through some of the northern Ontario communities around Sudbury. Today, scientists are still arguing whether this happened as a result of the greenhouse effect or for some other reason. We also have greater temperature swings than in the past. Preparation for rough weather starts off with knowing what to expect in your immediate area. The lack of preparations we make for annual recurrences of storms and blizzards is amazing.

There are different names for storms in different parts of the world. These are:
- Blizzards—Northern areas
- Cyclones—Indian Ocean and the Indian subcontinent

- Hurricanes—North Atlantic, Caribbean, North Pacific (east coast), South Pacific
- Typhoon—North Pacific (western), China Sea
- Willy-willy—Northwest Australia

Hurricanes have been called the "greatest storms on earth" because of their size and potential destruction. A hurricane can meld storm surge, powerful winds, tornadoes, and rainstorms into a devastating combination. As the climate changes, we will experience more and more hurricanes.

You should study the history of your local area with a particular emphasis on determining whether there is any periodicity to major storms and blizzards. Find out how your community coped with these events. This will give you your first clue to what is likely to come your way. Then look at your public services and find out how much curtailment of those services you will face after a natural disaster or a storm. The period of the expected disruption will establish your minimum self-sufficiency level. In practical terms, you should be able to heat your home, cook your meals, safely store your food, and look after personal hygiene needs for at least one week.

Day One

Tornadoes, blizzards, heavy snowstorms, and ice storms can isolate you from the rest of the community, leave you without power, or even cause you to get lost. Plan to seek shelter, get to know the area you travel, and have emergency shelter locations in mind while traveling.

It is very important to keep preparation in mind. The difference between being unprepared and prepared can mean the difference between a life-threatening event or a memorable day of making toast in the fireplace.

What to do:

- Have a small emergency kit handy. In winter, you should add additional candles and blankets to the kit. Be sure to have a non-electric can opener on hand.

- During tornado or storm season, have a portable radio tuned to a local station.

- Most tornadoes approach from the south or southwest, so check the sky from time to time.

- When you hear a hurricane warning, board up windows, and place tape over the inside glass. You should have plywood, shutters, plastic sheeting, and other materials on hand. Brace double entry and garage doors at the top and bottom.

- Always keep the car's gas tank at least half full.

- Keep a couple five-gallon containers full of water. Change the water every three months.

- Keep your home in good repair. Repair loose roofing and siding. Trim dead trees and tree limbs.

- Brace the gable ends of the roofs, and make other structural changes to improve the chances of your home surviving.

- Review your insurance policy. Check that you have both home-owner's and flood insurance.

- Have cash on hand. With no power, bank ATMs and credit cards will be useless.

- If you have warning, remove all loose objects from your yard, such as tricycles and trash cans.

- If you have a boat, either take it out of the water or secure it well.

- Leave your swimming pool filled and superchlorinated. Cover the filtration system.

Day Two

In the case of a tornado (a violently whirling wind that appears as a funnel-shaped cloud hanging from the base of a thunderstorm cloud), there is usually very little warning. Tornadoes are most destructive because of strong rotary winds, flying debris, and the partial vacuum formed in its center. The explosive pressure differ-ence can topple walls outward. Blizzards, snowstorms, and hurri-canes are usually tracked by weather-observation satellites, allowing some warning.

169

If you have an electrical outage, you can expect food to spoil and no air-conditioning or heating. Downed power lines or trees can damage your house and block roads.

What to do during a tornado:

- Stay away from windows, doors, and outside walls. Protect your head.
- Go down to the basement, or seek shelter under a stairway, a sturdy table, or in a closet.
- Try to reach the center of the house.
- If you are outside, lie flat in a ditch, ravine, or other depression. If you cannot find a depression, lie flat and hang onto a small tree or shrub.
- If you are in your car, stop and get out.
- Shut off electrical circuits if flooding threatens. Turn off major appliances like air conditioners and water heaters.
- Once you fill containers with water, shut off the main water valve.
- Doors and windows facing away from the approaching tornado should be left open to help reduce damage to the building.
- Store valuables in empty appliances—washers, dryers, dishwashers, or ovens.
- Put plastic bags over televisions, computers, and other electrical appliances. Place clothing in plastic bags.

What to do during storms or blizzards:

- Listen to your battery-operated or car radio for local broadcasts. Heed the warnings.
- Check for fires.
- If you suspect damage to your home, turn off the gas, power, and water. Use flashlights in case of gas leakage, and *do not smoke.*
- If the water is off, use other sources of water as explained in the previous section.

- If you have time, secure outdoor objects.
- Move away from beaches and low-lying areas.
- Check sewage lines before flushing toilets.
- If the power is off, use food from your freezer before it spoils. The freezer will protect food for forty-eight hours after the electricity is off, as long as the door is not opened and closed frequently. Take food out of the freezer only if it's absolutely necessary.
- Stay away from damaged areas.
- Stay away from downed power lines, fallen debris, and broken gas mains.
- Do not enter damaged buildings as walls may still collapse.
- Do not use the telephone except in a real emergency.
- Do not go sightseeing. Keep the roads clear for emergency vehicles.
- Check your food and water supplies before using them. Cans and other food containers should be examined for dents, rips, and contamination. Unless advised otherwise, all water should be filtered or boiled.
- If heating is off during a blizzard, stay in one room and warm it with candles. Be careful with fires. Someone should be awake at all times to watch the candles.

If you are trapped in your car in a snowstorm:
- Stay in your car. You won't get lost, and you'll have shelter.
- Keep fresh air in your car by opening a window.
- Run your motor sparingly. Beware of exhaust fumes, and make sure the exhaust pipe is not blocked by snow.
- Set out warning lights or flares.
- Exercise your legs, feet, and hands.
- Keep moving, and do not fall asleep.
- Watch for traffic or searchers.
- Keep in mind that a candle can keep the inside of a car above the freezing point.

Day Three

Check the damage done to your house, and take photographs if you can. You can make temporary repairs to keep out the elements. Duct tape and plastic sheeting will come in handy.

What to do:

- Report any broken power, gas, or sewage lines to the utility companies.
- Check with your insurance company.

Day Four

Put to work what you have learned from the storm. You may have found that some parts of your home are prone to flooding or other storm-related damage. See if you can improve on preparations before the next storm.

Lightning

The concept of lightning protection can be summed up in a few words: You, not Mother Nature, should have control over the lightning strike's energy. This involves providing a path to earth and not allowing the lightning to follow a random path. It is not possible to stop a strike, nor is it possible to prevent a strike from occurring. Therefore, you must be prepared to divert the strike energy via a deliberate and controlled path so that no damage will be incurred. The best diversion is with a lightning rod, but even a metal pole in the ground can serve this function.

Building or structure protection is more forgiving than protection of electronics. A building can handle 100,000 volts while electronic devices will be damaged with just a few volts.

If you are outside, the release of electrical charges built up in clouds can be especially dangerous on high ground or when you

are the tallest object. In a lightning storm, keep away from hilltops, tall trees, and boulders. Make for low ground, and lie flat.

If you cannot get away from tall objects, but have insulating clothing like rubber-soled boots, sit it out. You must have dry material to insulate you. Do not sit on anything wet. If you cannot insulate yourself, lie flat.

You can usually sense that lightning is about to strike. The signs are a tingling of the skin and the feeling that your hair is standing up. When you have this feeling, lie down quickly. Get away from large metal objects.

If you can get in a cave, do it. In a cave or building, stay at least four feet away from the walls. Stay away from doors, cave openings, or overhanging rocks.

Floods

There is nothing like waking up at night to find the bed floating around your room. It is even more disconcerting if this happens without warning. If you live on a flood plain, you should unload your property and move somewhere else now. I know it is hard to do if you spent ten years scrimping to save up the down payment on your house, but in the U.S., some 50 million acres flood annually.

Most floods are at least somewhat predictable. However, that is small comfort when the waters enter your home. The floods along the Pacific Northwest, the Rocky Mountain, and Great Basin areas occur in certain seasons. On the other hand, floods along the Gulf and southeastern coasts are well nigh unpredictable.

Dams can be another source of flooding in your area. The most dangerous time is when the dam is filling for the first time. If you are downstream from such a dam, it may be wise to visit friends in higher areas, especially when a storm is brewing.

Floods bring with them the risk of infectious diseases and provide a breeding ground for disease-carrying insects.

Day One

Day One is not the day of the flood. Day One is before the flood, when you have time to prepare. For example, serious danger of flooding in the lower reaches of a river occurs when flood crests from tributaries arrive at the same time. Another cause is when an ice jam blocks the flow of a river, or in a fast-running river, the ice submerges to form "frazzle" ice. Get ready first. Then when you hear reports of cresting and ice jams, you won't be frantically running around doing whatever you can rather than whatever you should.

If you see discoloration on the basement walls of a house you plan to buy, this is a warning sign to look into the flood history of the neighborhood.

What to do:
- Have a small bug-out kit prepared at all times.
- Find out if your property is above flood level and how many feet you are above a major flood.
- Have a portable radio tuned to local broadcasts.
- Keep your vehicle fueled and ready to go.
- Have plastic sheeting, sandbags, and lumber for waterproofing your home.
- Purchase a battery-powered sump pump for your basement if your house needs a sump.
- Have your immunizations up to date.
- Have activated lime on hand for disinfecting carcasses.
- Identify more than one way to leave your area. Note any low spots prone to flooding, and find ways around them. You should drive your escape roads to become familiar with landmarks and potential hazards.
- Purchase a weather-alert radio.

Day Two

There is usually a warning in the case of a flood. If you are asked to evacuate, do not try to ride it out, or your face may be on CNN as the helicopter tries to pluck you off your roof. As soon as you hear

the warning, prepare to evacuate. If you live in a flood-prone area, all of your goods should be highly portable.

What to do if you are home:

- Turn off basement furnaces and the outside gas valve.
- Shut off electricity. If the area around the fuse box or circuit breaker is wet, stand on a dry board and shut off the power with a dry wooden stick.
- Never try to cross a flooded area on foot. The fast water could sweep you away.
- Keep plenty of drinking water in closed containers, and have plenty of canned juices.
- Have ready-to-eat foods on hand.
- Move furniture and electrical appliances to a higher level in your home, if possible.
- If your home is likely to flood, move to a safe area now.
- Never cross any flowing stream having more than one foot of water or more or one that carries pieces of wood or ice.

What to do if you are in a car:

- Drive very carefully.
- If the car stalls in a flooded area, abandon it. Many people have drowned in rising floodwaters while trying to move a stalled vehicle.
- Do not attempt to drive through a flooded road. You can be trapped by rising waters.

Day Three

The day after the flood. The place may be a mess, and you may hear warnings to boil water. Dead animals and overflowing septic fields may cause the outbreak of contagious diseases.

What to do:

- Assess the damage, and call your insurance company.

- If the power has been off for more than forty-eight hours, you may have to discard the contents of your freezer and refrigerator.
- Wash all affected areas with disinfectant.
- Canned goods may be used if you inspect and clean the cans. Remove the labels, and wash the cans with soapy water.
- Check your food and water supplies. Do not use any fresh food that was in contact with floodwaters. Check food in your freezer and fridge.
- Wells should be pumped out and the water tested before using. You should place bleach in your well overnight. The next day flush all water outlets until the heavy chlorine taste is gone.
- Do not handle any live electrical equipment or lines. All electrical equipment should be completely dried out before using.
- Don't go sightseeing. Flood victims and emergency personnel have enough to do without visitors.
- To examine the basement or other buildings, use flashlights, not lanterns. There is a chance fuel from containers or broken lines may be present.
- You should not handle dead animals. Cover them with lime and a layer of earth as a temporary measure.

Day Four

Authorities will announce new flood-control measures. These can range from adding dikes to evacuating residents into less flood-prone areas. If there are government aids available for moving from a flood-prone area, take advantage of them and move.

What to do:

- Report any broken power, water, or gas lines to the respective utility.
- You will need to drain any pools of floodwater on your property. Remove all dead animals very carefully, as they may be a source of typhoid.

- Be sure to photograph all damage done to your home before cleaning up. Prepare detailed lists of items damaged in the flood. The lists should include serial numbers, if any, date of purchase, name of the seller, and the original purchase price.

Day Five

By now you should decide whether to move or stay put. Sometimes it is better to take a loss than to go through the frightening experience of a flood all over again. Do not sell a property right after a flood, though. Fix it up, and wait for a better time to unload it.

What to do:
- Review with knowledgeable people the likelihood and frequency of floods.

Day Six

Find out if the flood that affected you was a once-in-a-century or once-in-ten-years occurrence. Act accordingly.

Landslides

Population growth has caused a push into marginal areas. There, land development encounters landslide hazards and, in the process of building roads, creates new ones. Just look at the Santa Monica and San Gabriel Mountains close to Los Angeles. One day there are grass fires, and the next day it is a landslide. Sometimes the two go hand in hand. Landslides and mudslides are ever-present dangers in hilly and mountainous areas.

A landslide can start because of slow erosion, heavy rains, earthquakes, engineering design defects, or glacial action. In the past, whole communities have been wiped out by landslides. People living on sloping ground are vulnerable to landslides. Everything in the path of one is destroyed. Another form of landslide is a sinkhole,

which can form when mining activity, oil recovery, or an underground river carve out large cavities. Sometimes these collapse, and whole villages are swallowed up. We hear about landslides after major rains.

If a landslide blocks a river, we are faced with the additional problem of flooding. You should evaluate all possible risks to your property prior to any disaster.

Day One

The best way to avoid being caught in a landslide is to keep away from slide-prone areas or travel through them when the likelihood of a slide is minimal. Most likely times for a slide would be during or after a heavy rainfall, during an earthquake, construction, the spring runoff, or after forest fires.

What to do:

- Choose building sites with care. Do not locate on a slide path or in a ravine. Before you build, have a soil engineer assess the property's probability of being in a landslide.
- When camping or hiking, avoid ravines and sheer cliffs. Any hillside lacking vegetation can present a slide hazard, particularly if it is moisture laden.
- Before buying or renting a house, check the basement for cracks in the floors or walls. In the neighborhood, tilted power poles and displaced street asphalt are signs of earth movement.

Day Two

Your building or your area was subject to a landslide. You survived, so what do you do now? Take a cold, hard look at whether it is likely to happen again. You may want a soil engineer to look at your domain to see if additional slides are probable.

What to do:

- If you see or hear a slide coming, run for it. Run at an angle, downslope and away from the landslide.

- Don't rush into an area where a landslide has occurred. Sometimes additional landslides follow the first one.
- If you are driving through a landslide area, be careful of debris-laden sections of the road. It is better to turn around and find a safer route. Better late than dead.

Day Three

Now that the slide is over, assess the damage.

What to do:

- If your building is intact, check for broken gas, water, and sewer lines, as well as fallen electric wires. If you find any, report it to your local utility.
- Take photographs of damage, and make a list of damaged or destroyed possessions.

Day Four

Time to take stock. You should contact your insurance company if there is any damage.

Avalanches

Most avalanches occur in sparsely populated areas. I guess that is why they are sparsely populated. The "loose-snow" avalanches start at a point, much like a snowball rolling downhill, and consist mostly of loose snow. In contrast, a "slabavalanche" starts when a large area of the snow starts to slide. The mining camp at Wallace, Idaho, was wiped out by an avalanche in 1906. In the same year Mace, Colorado, was hit—more than 120 dead. The terrain most likely to have an avalanche will have either a short or long slope and large rocks and trees that tend to anchor snow.

Although avalanches and landslides have much in common, there are major differences. For example, you can use vegetation as

a rough guide on slopes to see whether the area is avalanche prone. The table below is from the U.S. Forest Service.

Avalanches are endemic in some areas of the country, so learn about the history of the area you live and travel in.

Day One

Avoid avalanche-prone areas. It is better to detour around the area than have to use the desperation measures detailed in this scenario. However, if you are caught in an avalanche, it is best to know what to do.

What to do:
- Avoid north-facing slopes in winter.
- Smooth and grassy slopes are avalanche prone in the winter.
- Keep noise level to a minimum.
- Wear a locator beacon.

Vegetation as a rough indicator of avalanche frequency	
Vegetation clues	Avalanche frequency interval
Bare patches, willows and shrubs, no trees higher than about 3 to 6 feet.	1–2 years
Few trees higher than 3–6 feet. Immature trees of disaster or pioneer species. Broken timber.	2–10 years
Predominantly pioneer species, young trees of the local climax species (increment core data).	10–25 years
Mature trees of pioneer species, young trees of local climax species (increment core data).	25–100 years
Increment core data.	Over 100 years

Day Two

This is the day of the avalanche. Sometimes there is warning that avalanche conditions exist, but most people will ignore such warnings. Avalanches take lives because people take the attitude that it can't happen to them.

What to do:

- Discard all equipment, and jump clear of a snowmobile if you're on one.
- Make swimming motions. Stay on top if you can. Work your way to the side.
- As the slide slows, make a guard for your face with your hands. This will create an air pocket.

Day Three

Now that the avalanche is over, it is time to look for survivors.

What to do:

- If you have report of people buried, form a search party.
- If the ones buried wear locator beacons, your job will be easier.

Day Four

The aftermath of an avalanche can be a useful period. Get advice on whether a recurrence is a periodic event or if it was a once-in-a-lifetime occurrence at that location.

Volcanoes

Volcanoes can be extremely destructive, but thankfully, most of them are located in sparsely populated regions. Volcanic eruptions are rare events. Volcanic activity takes many forms. Geysers, hot

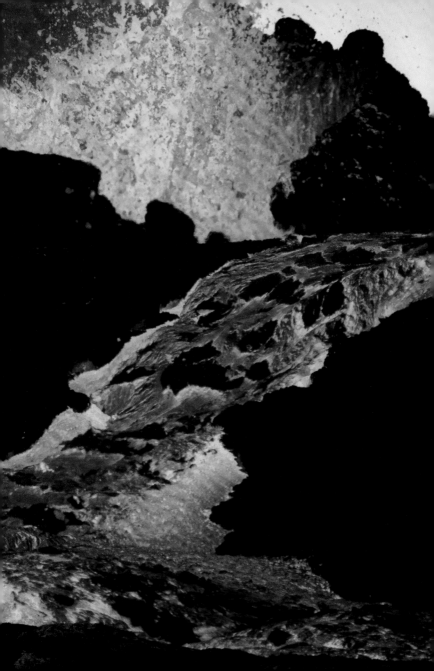

springs, and gas emissions from vents and fissures are as much part of the volcanic activity as are lava flows and emissions of ash into the atmosphere.

In addition, four types of volcanoes exist. These are the cinder cones, composite volcanoes, lava domes, and shield volcanoes. Lava domes can be most destructive. A global cooling is associated with large-scale volcanic activity. Most of this is due to the formation of sulfates in the stratosphere. These, along with the other crud thrown up, reflect sunlight back to space. The stratospheric sulfates are tiny and reach high elevations in our atmosphere.

Volcanoes have caused great damage and loss of life. The AD 79 eruption of Vesuvius buried Pompeii and killed 16,000. The 1906 Mount Pelée eruption in Martinique destroyed the city of St. Pierre and killed 30,000 people. Eruptions are some of nature's most powerful events. They have been with us since the formation of our planet, but we are presently in a quiet phase of volcanic activity. (But don't try telling that to the citizens of Montserrat.)

Before the lava flow, there is usually an ejection of cinders and old lava fragments. This clears the crater for the eruption of lava. After the cinders fly, there is usually a short period of quiet before the lava erupts.

Most of the potentially active volcanoes in the U.S. are sited in Hawaii and the Cascade mountain crest, covering Washington, Oregon, and California. There are others that were active in historical times.

Day One

Increased volcanic activity may suggest an eruption in the near future. There may be official warnings. The following eruption hazards are the ones most likely to face you:

Lava—You can outrun most basalt lava flows. However, they will run until they reach a valley or similar low place. Only then will they cool off.

Missiles—These range from pebble-size fragments to large rocks. These bombs can fly great distances from the eruption.

Ash—Volcanic ash is pulverized rock forced out in a stream of steam and can cover greater distances than missiles. It can cause roofs to collapse. The ash smothers crops, blocks transportation avenues, and causes lung damage to people. The ash is the mother of all acid rains.

Gas balls—These can run down the side of a volcano at a speed of 100 miles per hour or more. Your only chance is to be underwater when one reaches you.

Mud flows—There's nothing like a flood to get your attention. In a valley, this can have the most destructive effect of all. If you see a mud flow coming at you, run diagonally up the side of a hill or away from the mud flow.

What to do:

- Have a bug-out kit ready to go.
- Keep abreast of local and national news.
- If you live in a volcanically active area, you should have a neighborhood warning network.

Day Two

This is the day of the eruption. Undersea volcanoes usually have a nonexplosive lava flow, but those on dry land are usually accompanied by explosive lava flows. One side effect of the lava flow is large-scale disruption in land use. Decomposed volcanic material makes excellent farmland, but this will take centuries. In the meantime, the land is useless.

Most volcanoes give plenty of notice of a potential eruption. The science of predicting eruptions is much better than in the past.

What to do:

- Go for the high ground. Lava tends to follow riverbeds, valleys, and roads.
- If the ash starts to fall, close all windows and fireplace dampers. Wear a painter's dust mask or a wet cloth. It is very important to protect your eyes. A gas mask is very good if you have one,

but a motorcycle helmet with a visor or a scuba diving mask will do.

- Acid rain falls with the ash. If you're caught in the open, upon entering a shelter, discard clothing and wash all exposed areas with water. Flush all exposed areas repeatedly.
- If you have to evacuate, keep car windows closed to keep acid rain from getting inside.

Day Three

The initial eruption is over. Now the question remains whether further activity will follow. Volcanoes will vent gases and steam for some time after an eruption. If there is no venting, the possibility of a plug in the volcano exists. This may lead to another major eruption.

What to do:
- Keep tuned to the news reports.
- Take photographs of the damage, and list damaged items.
- Stay indoors if further eruption is expected.

Day Four

All is quiet now, so take stock of the damage. Remove volcanic ash from the roof of your house and from your property. You may have to remove more than three inches of the topsoil to rid your property of the ash.

What to do:
- If your home has been damaged, find out about government programs available for the affected areas.
- Listen to public announcements to see if there are further eruptions.

Day Five

Time to see whether the volcanic event was isolated or if it is a prelude of things to come.

W hat is the difference between a terrorist and a freedom fighter?

A freedom fighter chooses a legitimate target. Freedom fighters do not massacre men, women, and children. Terrorists do. Freedom fighters don't blow up restaurants. Terrorists do. These are only a few of the differences between terrorists and freedom fighters. There are others, but these relate to the details of the conflict. You will find these in the following scenarios.

Terrorism is big business nowadays. We have all flavors of it. The religious right, the radical left, and the dislocated minorities are just some examples of potential terrorist threats. Bombs are very bad soldiers. They do not distinguish between friendly and enemy units and kill indiscriminately. You cannot even surrender. While governments call bombs the weapons of cowards, the practitioners of this type of warfare call it the "poor man's revenge."

However, bombs are only a small part of the terrorists' arsenal. There are other methods they use to create havoc, such as throwing a diseased animal carcass into a city's water reservoir, letting natural gas accumulate in an empty house and using a remote-controlled spark generator to ignite the gas, and blowing up electric grid towers.

The world has changed incredibly. *We can never feel totally secure anymore.* Once you accept that premise then you can intelligently think about how to reduce your chances of becoming an unwitting target of terrorism. Preparation starts by looking at your routine activities. List them and the routes you take. Once you do this, check your routes for chokepoints (obvious ambush sites) and whether your route takes you past any potential targets. Does your route go by a state or federal prison housing convicted terrorists? If you cannot avoid passing it, lock your car doors, drive defensively, and make sure you have enough gas to get you to your destination. You should make a habit of topping off your tank when it is half full. This is a prudent precaution applicable to all scenarios.

We live in an age of unprecedented changes. One of these is the ability of an individual or a small group of people to wreak havoc on the population. This ability, coupled with festering sores of nationalism and elitism, forces us to consider how to protect ourselves and our families from the effects of these new threats. These threats come in all types and sizes, but we are not completely defenseless.

This chapter was written for civilians, not unlike the ones who were victims of nerve-gas attacks in Tokyo. We are the ones who have neither fancy isolation suits nor digital survey meters. We must get by with the supplies on hand should our local neighborhood terrorist decide to engage in some low- or high-technology mayhem.

Sarin nerve gas used by a religious cult in the Tokyo subway. Chlorine gas spills from an overturned rail car. Fuel tanker hitting an overpass. Terrorists obtaining nuclear materials. These are the headlines in our newspapers today. And don't forget state-sanctioned terrorism.

The question of a terrorist's use of nuclear, biological, and chemical (NBC) materials was always a question of *when* rather than *if*. What now? There is a whole new class of terrorists out there, well-educated and many with professional backgrounds. This is a new trend. Airplane hijackings, hostage taking, and the like were once relegated to former U.S.S.R. lands. Terrorists there are on the way up on the learning curve. In the meantime, we on the North American continent must be ready to deal with new threats to our well-being.

At one time, with the two superpowers facing each other, nuclear survival measures were simpler. When the saber rattling escalated, you "bugged out." That is to say, you evacuated from the target areas. Now we have no visible escalation ladder to provide a warning time; we have no forewarning. While the threat is at a lower level, it is much more unpredictable. In the past many knowledgeable people moved out of cities and took up a more pastoral lifestyle or moved to smaller cities and towns. Is this still a good protective measure?

Let us go back to the newspaper headlines:

* Amtrak train derailed in the Arizona desert
* Bombing of federal building in Oklahoma City
* Fringe group terrorizing small town
* Rail car derailment forces evacuation of village
* Atlanta Olympic park bombing
* Hijacked jets destroy Twin Towers and hit Pentagon in day of terror

It seems that the modern-day terrorists have moved also, perhaps in search of a safe house, a chemical factory, or a training base. Now the terrorists are out in the villages and the countryside, among those who, in search of security, fled the major cities. This forces us to take another look at protection. No place is safe anymore. Therefore, we must be aware of what countermeasures can be taken to protect against the effects of NBC or other forms of terrorism, and we must be able to take them wherever we are.

Where to go? What to do? These are the questions facing a person confronted with the possibility of high-tech terrorists in Mr. Rogers's neighborhood. There is no place to go, but there is plenty you can do to protect family and friends. First and foremost, have a plan of action. This plan must be made during the cruel light of day, facing up to the unappetizing choices available to us. You must have information, which can come from your scanner radio, your radiation-detection instruments, or by word of mouth. But information you *must* have. This is vital.

Without information, you will be preparing for events not even remotely connected to what is happening in your area. You must also have supplies on hand to wait out the danger period, which can range from a few hours to three weeks.

What are these supplies? An abbreviated list is shown below:

Sleeping bag	Cooking utensils	Stove or fireplace
Knives, forks, spoons	Kerosene or electric lamp	Can openers
Cups and plates	Flashlight	Garbage bags
Matches	Crowbar	Shovel
Fire extinguisher	Soaps and detergents	Towels
First aid kit	Aspirin	Food
Water	Swiss Army knife	Detection instruments
Self-protection devices	Radio, multiwave	Extra clothing
Batteries	Maps	Compass
Gas mask	Rubberized suit	Decontamination agents

Once you have these supplies, you must have the knowledge to use them. The following scenarios should give you an introduction to the knowledge you must have to cope with the situation. They are not pleasant. And in a perfect world, we wouldn't have to deal with them, much less think about them. But what is perfect anymore? These scenarios have migrated from the battlefield to the backyards of our nation. We cannot ignore them, for the powers that should protect us are completely unprepared to protect even themselves. As always, the individual must look after him or herself.

The preparation has its own timetable, but first you must evaluate your neighborhood and area. To start, ask the following questions:

- Am I living close to a transportation hub?
- Is my area near an international port?
- Would my area be a nuclear target?
- Is my area home to prominent people who may be targeted by terrorists?

The answer is probably yes to one or more of these questions. Now what? Now is the time to take out the map of your area and decide what to do. First you must decide whether to stay where you live or evacuate. This is a very important decision. Evacuation is a must if your area is subject to a nuclear blast. If you are faced with biological and chemical threats, you may be better off staying put.

Let me elaborate. Unless you have a five-ton truck, clear roads, and the fuel to get you to your destination, you may be leaving behind equipment and supplies that will make your existence easier. Most evacuations in the case of terrorist acts and chemical spills are for a short duration only. In the meantime, your possessions are at the mercy of the authorities, however well-intentioned they may be. To top it off, the authorities will probably have roadblocks and will try to channel the movement of people into other areas. These areas will have so-called reception centers.

Another sobering thought: We worry about terrorists, but the largest toll of life comes from accidents connected with big business and the state. The Tokyo subway incident took nine lives. Bhopal took thousands. So did Chernobyl. Enough said.

Day One

There are terrorist threats and actions all around us. Therefore, it is very difficult to tell when random events turn into an organized mayhem that threatens your way of life. The government generally has more information than the average citizen. When you see the government restrict vehicular access to certain buildings, when you are searched upon visiting a government office, when there are guards around water reservoirs, and when other measures come into force, you know that there is a definite threat out there. Rule number one for self-preservation is *never become predictable!*

Why is that? Well, imagine for a moment you are predictable. A terrorist can enter your residence, place a pound of black powder in your microwave oven, and set the timer for three minutes after your usual arrival time. Or take the lightbulb out of your favorite reading lamp, file a small hole in it, fill it with black powder, and stick a piece of tape over it. When you turn the light on to read your newspaper, you will have a place in the next day's obituaries. Or the terrorist might look under your sink and read the warning labels on the contents. Mix vinegar and bleach, and presto, chlorine gas.

Architects are learning to cope with terrorist threats. Some of the steps they are taking include:

- Eight-inch base slabs with the supporting columns going all the way down to the foundation
- Tiered buildings with tiered landscaping, to reduce the area subject to blast
- Rebar in floors and ceiling laid in a crisscross pattern, and the rebar interconnecting with the rebar in the columns
- No external ornaments that can fall off
- Blast curtains and mylar coating on the insides of all exterior windows
- Reduced window areas to 20 percent or less and spacing columns closer together (twenty-five to thirty feet)
- All construction poured-in-place concrete with twelve-inch exterior walls and eight-inch thick roof slabs
- Even the employee parking at least 100 feet from the building and no underground parking
- In effect, the building is turned into a fort.

What to do:
- Take a defensive driving course.
- Do not drive by yourself.

- You can make good money by purchasing shares in companies manufacturing surveillance equipment.
- It is best to conduct your affairs as if a terrorist incident could happen to you. Keep your eyes open for people following you, strangers in your neighborhood, and other unusual events.
- Install a home security system.
- Have a safe room in your house with a steel door, a telephone line in the room, and some ready-to-eat food and water.
- Keep abreast of local and international news.

Day Two

By Day Two, the nature of the terrorist threats is clearly identifiable. Security will be enhanced. Turn to the section that applies to your situation and read on.

The External Threat

Terrorism coming from other countries is the poor man's way of waging war on a powerful enemy. The Irish attacks in Britain are good examples of this type of terrorism. We've had some examples on this side of the Atlantic. The World Trade Center is but one of them. Offshore terrorist organizations send their members into the U.S. as tourists, students, or whatever. Once here, they go underground and establish safe houses for others, often illegals, to operate from. Other external threats were demonstrated in Saudi Arabia, Lebanon, Germany, France, and even on the high seas. Instead of the "fog of war," we now have the "smog of terrorism."

If you are an American traveling in foreign lands, you must be careful. How careful? Careful enough to carry two wallets, one to give to the hijackers of your plane. The "giveaway" wallet should have your driver's license, your library card, and like items. Your second wallet should contain your armed forces or government employee identification card, or any other documentation that would select you out for *special* treatment by your captors. This is just the first precaution you must take.

Additionally, there are the terrorist sympathizers. They are not active terrorists, but because of political beliefs, ethnic background, or other reasons, they will provide aid to the terrorists. These people are very difficult to identify. They may be Americans and unless they have a terrorist history, will be practically invisible as far as counterterrorist efforts go.

The external terrorists are a means of carrying guerrilla warfare to the enemy's homeland. We have many enemies, most of whom are impotent to attack a superpower head-on. As a consequence, they directly or indirectly engage in supporting terrorists to carry the war to the U.S.

Day One

Yes, America has enemies. Any powerful world power is bound to have them. Because we are a superpower, these enemies do not dare to attack us openly. They resort to terrorist actions. Day One is here and has been for some time. What is surprising is that we have had relatively few incidents so far.

These terrorist acts are becoming more serious and threatening, partially because of the complexity of our civilization and partially because of the wide availability of indirect weapons. In the 1950s, a sabotage on a refinery may have put the refinery at risk. In today's world, not only is the refinery at risk, but the adjacent densely populated suburb as well.

The Border Patrol is coming under increasing threats. At one time, they had to deal with economic migrants in search of jobs. Now there are shooting targets tacked onto the fence along the U.S./Mexico border. The targets have an officer's silhouette and many holes shot through them. Then there are the very sophisticated transmitters confiscated in the southwest portion of the border. Whether this is associated with drug runners or foreign terrorists is irrelevant. Both are aimed at penetrating the United States.

What to do:
- As stated repeatedly, have a bug-out kit.
- Review the identification you carry with you. It may be prudent to leave home your twenty-year-old Marine Corps sharp-shooting certificate and other similar identification papers.
- Conduct a potential target evaluation. Know in which countries your employer operates.

Day Two

When you hear of increased terrorist activities following an international incident, then you should try to terrorist-proof your home, vehicle, and workplace. The precautions do not have to be so restricting that you become a prisoner in your home, but if you happily continue as you live today, you may be inviting an unplanned kidnapping or other terrorist act.

Islamic extremists pose an increasing terrorist threat. International terrorists do not congregate exclusively in New York and Washington. They are in rural settings as well, which increases the problems facing the agencies engaged in the fight against the terrorists.

What to do:
- Fly El Al, the Israeli airline, if you can. Their preflight and in-flight security is unmatched to date.
- Avoid large gatherings if you can. Any well-advertised event may also attract unwelcome attention by terrorists.
- Do not use self-storage lockers. Close out any you may have. There are other storage options, such as private garage rentals.

THE SURVIVALIST'S HANDBOOK

Day Three

At about this time, even the government realizes that we are under attack by foreign-based terrorists. This will result in persecution of those nationals living on American soil, thus providing a fertile recruiting ground for the real terrorists.

The public sentiment against the terrorists in the U.S. will have an international impact. In some countries, people will make life miserable for Americans traveling there. Americans working for international aid agencies may be targeted by terrorist sympathizers. U.S.-owned businesses in foreign countries will be targets of local sympathizers of terrorists and wannabe terrorists.

Many underground publications advocate the use of self-storage lockers to store arms, ammunition, explosives, drugs, and other substances. Police departments will set up surveillance and raid many of these self-storage establishments. Those remaining in operation will insist on verification of the identity of the would-be renter.

What to do:
- Take a good look at the makeup of your neighborhood. If you find that a large number of people are nationals of the country conducting the terrorist raids, you may want to move or increase your insurance coverage.
- Listen to shortwave broadcasts coming from the country exporting the terrorists. Find out if they have any specific cities or events they may want to target.

Day Four

The U.S. government will take steps to teach the country supporting the terrorists a lesson. The reprisals may or may not work, depending on the sponsor country's nature. The terrorists may have religious as well as national overtones. This can

turn the fight against terrorism really ugly. The U.S. will impose economic sanctions against the host country of the terrorists. The UN will be livid and will try to censure the U.S. on the basis of discrimination.

What to do:
- Review your investments. If any of the companies you own stocks in have assets in the target countries, it may be time to sell them.
- Invest in security-related stocks.
- Find out which items are imported from the country behind the attacks. It may prove profitable to stock up on those items. This has to be done in a logical fashion—Iranian rugs are not all that popular nowadays.

Day Five

The terrorists will step up their activities in the U.S. They will target government installations, hydroelectric dams, power transmission lines, rail yards, places where people gather, and anywhere where they can make the headlines.

It is likely that U.S. armed forces will make a demonstration attack on the host country as a form of out-of-court settlement.

What to do:
- Plan your routes to avoid targets.
- Drive with your doors locked and the windows up.
- Start a security company. Business will be good.

Day Six

The terrorists will intensify their activities, now hitting any place where people congregate. There will be a number of symbolic targets, such as the Vietnam War Memorial if the terrorists are of Vietnamese origin. It is quite likely that local vigilantes will make attacks on neighborhoods with large concentrations of people originating from the terrorist country. This will result in widespread random violence.

The Immigration and Naturalization Service will make raids on ethnic neighborhoods. These will result in the dispersal of people from the target area to other parts of cities. In some cases, this will lead to violence.

What to do:

- Take a good look at where you live and travel. Identify potential targets, and avoid them.
- Avoid public gatherings, particularly those associated with the cause of the terrorists.

Day Seven

Around this time you will see whether the government's efforts to eliminate the terrorists are working. If they are not, the many so-called temporary restrictions will become permanent. If the terrorists are being eliminated, these restrictions may be relaxed or lifted.

No matter which way it goes, the events will reinforce in the American psyche a mistrust of foreigners. This mistrust may create another terrorist movement among the legal immigrant population. It is likely that local police forces will also have to check on immigration status of people in their district.

What to do:

- Make sure that your permits are up to date. Law enforcement is likely to be very rigorous.
- If the terrorist threat continues, set up several security guard companies. Private security business will be brisk.

Day Eight

One of two things will happen: Either the nation will be an armed camp with travel restrictions permanently in place, or the terrorist

threat will have been eliminated. The terrorists may have gained the support of expatriates living in America, who will provide safe houses for them. This will make them very hard to eliminate.

What to do:
- If the terrorist activities continue, try to work from your home. This will eliminate much commuting and reduce your chances of being an unwitting target.

Day Nine

Continuing terrorism will lead to a complete elimination of our freedoms, so the terrorists will have won.

The Internal Threat

Seeing the Oklahoma City's Alfred P. Murrah Federal Building blasted into rubble served as a wake-up call to all of us. Domestic terrorism is a fact of life now. Previous terrorist acts went almost unnoticed, but the loss of life in blowing up the federal building could not be ignored. More recent examples include the anthrax letters, the beltway sniper attacks, and the shooter at the Holocaust Memorial Museum. The American scene is replete with frustrated groups giving up on the judicial process. They are ready to take the issue directly to the so-called oppressors. In the past, we dealt with the perpetrators solely through the criminal justice system. Now it is more complex.

Part of the problem comes from the increasing powers of the federal and state governments. The old-time eccentrics now have become potential terrorists. It seems that the more pressure is put on the population to conform, the less nonconformist behavior the authorities can tolerate—a rude reminder of the second law of thermodynamics.

Mind you, we have had internal terrorism before. Back in 1977, a Senate judiciary committee held secret hearings looking into sabotage attempts on pipelines, plans for hijacking an offshore oil platform, and an attack on a refinery. These terrorist activities were hushed up by the authorities to discourage copycats. Decades later, our contingency planning to deal with terrorists is still inadequate. Except for some efforts by the FBI, we are reactive rather than proactive. That is, we will want to shoot the horse after it has bolted from the barn.

Because of the pressure on law enforcement agencies to control terrorism, police intelligence agencies will keep files on all types of people. These terrorist profiles will be all encompassing. For example, undercover officers are regularly fingered as terrorists by the terrorist profiles used. To compile these profiles, the government will use all sorts of databases. What might get you on the list would be the purchase of four SKS carbines from your local gun dealer and filling out of the BATF forms, the phone call you made to Aunt Lucy while she was in Lebanon, the Visa charge at Wal-Mart for 500 rounds of ammunition, the credit card receipt for the fifty pounds of garden fertilizer you got at Joe's Garden Supplies, your membership in the NRA, your subscription to the *New American*, the fax you sent to the German government protesting the shipment of nuclear waste to France for reprocessing—and these are just a few. If you knew how much information the National Security Agency has on you, you would have an unrelenting headache.

There is also terrorism by groups and organizations for animal rights, nuclear disarmament, and other similar causes. Their tactics of direct action include tree spiking, pelt painting, product contamination, spring-loaded center punches on windows, glass-etching fluids, rooftop demos, breaking and entering, masking scents against hunting, damage to vehicles, bringing false charges against people, "ethical" theft, and "ethical" shoplifting. (Of course, "ethical" theft and shoplifting are nothing but self-justification for the criminal acts. Stealing from a mink farm is still stealing.) These are mostly small-scale, sporadic events, but they can be threatening to people living in the target areas.

Day One

Day One came during the past several years. You can recognize Day One by not being able to park in the vicinity of a government building, by needing an escort to various offices, and by the searches of your briefcase. These are standard operating procedures in Ireland and other places beset by an internal terrorist campaign. Security will be beefed up at other locations as well, such as corporate offices and factories.

If you work for a large organization, it probably ran a security check on you. Its subcontractors will have their employees' backgrounds examined.

What to do:
- Have a bug-out kit.
- Make arrangements to reduce your need to visit government offices and installations.

Day Two

The government will give itself unprecedented powers for the "duration of the emergency." Phone taps without court orders and arrest and questioning of suspects without due process of the law are just two examples. Many people feel that this will be a vehicle for the "New World Order" to take over America. Time will tell.

Fertilizers, such as ammonium nitrate and sulfur, will have special coating substances added to reduce their usefulness for making explosives. There will be input-output balance forms to ensure that the purchasers do not divert their fertilizers into making explosives. Gunpowder may contain "taggants" to identify the source of the material. Sales of certain other chemicals will be severely restricted, and local police will follow up on local purchases to ensure that these are used for legitimate purposes. Woe be to you if you are

a drag racer or a model-rocket enthusiast. Of course, the terrorist will happily use a solvent to remove the coatings from controlled substances, steal the supplies he needs, and continue as before.

What to do:

- Prepare a checklist of the organizations you belong to, your past associates, and your past purchases to evaluate if you would be on any of the governmental "special" lists.
- If you are on a government list, try to change your address, and, if possible, get a new identity.
- Do not joke about bombs or any other sensitive subject. You will find yourself under police scrutiny even for a silly joke.

Day Three

More government restrictions will be placed on the population in the name of law and order. As more restrictions are imposed, more resistance will be put up by the population at large. This will lead to even more power being placed in the hands of the authorities. Of course, these will be strictly "temporary" in nature. Don't you believe it. Look at the introduction of the social security number. First it was only for social security. Now you can't even open a bank account without providing it.

What to do:

- Form several corporations. These will enable you to bank without social security numbers.
- Rent safety deposit boxes where you bank for your corporate papers. You won't have the papers around in case you are raided and will be able to keep part of your activities private.

Day Four

The random violence will increase. There will be bombs exploding at public gatherings, and bomb threats will disrupt life in general. The authorities will respond with even more controls to reduce access to government facilities.

We may see the emergence of a secret government. As science fiction author Henry Kuttner wrote, "Up till now, no race ever successfully conquered and ruled another. The underdog could revolt or absorb. If you know that you are being ruled, *then the ruler is vulnerable.* But if the world doesn't know—and it doesn't."

What to do:
- Keep your visits to federal and state offices to a minimum. Use the mail, phone, and faxes to deal with government agencies.
- Consider setting up a company dealing in security and surveillance devices.

Day Five

About this time, you will need a travel permit to visit Washington and other major target cities. Border areas may require a special permit to visit, and the use of exit visas may be introduced to control who is leaving the U.S.

Many constitutional guarantees will be suspended. People will be held without charges. Preventive detention will enable those with grudges to get even by turning people in on trumped-up charges.

What to do:
- If you find that terrorist activities increase, buy your supplies wholesale to reduce the number of trips you must make.
- A cooperative neighborhood watch should be formed.

Day Six

The list of proscribed goods will increase. You may find that your gardening activities will suffer due to the shortage of fertilizers. Moreover, making your own may result in bomb-making charges being brought against you. If you were considered an eccentric before, now you may be classified as a potential terrorist. Many innocents will be questioned and treated as terrorists.

Roadblocks on major roads will be erected more and more frequently. They may become a permanent fixture on some major roads leading to large cities.

What to do:

- Take an inventory in your garage and your basement. If you have short pieces of pipe, threaded end caps, ammonium nitrate fertilizer, sulfur, and other items suitable for making bombs, you'd better cache them now.
- If you are worried about your profile, establish several caches of firearms and supplies in case you have to bug out.

Day Seven

The authorities will try to curtail terrorist activities by profiling even more people. The criteria will be so loose that anyone with a grudge will be able to turn in another person. The neighborhood snitch program will be in full bloom. These measures will not reduce the terrorist threat, since by now any eccentric will know that he is targeted by the authorities. The government's actions will serve only to increase unrest. There will be widespread acts of minor sabotage aimed at the government, which in turn will spawn even more repressive actions. The government unwittingly will have created a self-fulfilling prophecy.

Due to these repressive measures, many people will be driven into the arms of the terrorists. What started as a terrorist movement may become a liberation movement. That is, liberation from a police state and government tyranny.

What to do:

- If you fit a profile, change your name and move.
- Identify the neighborhood snitches and have a midnight discussion with them explaining your new health program. If they don't listen, get rid of them. This is not the time to be timid.

- Be very careful what you communicate over the Internet, telephone, or any other communication medium. These will be monitored.

Day Eight

Unless terrorism is brought under control, we will see the emergence of a police state. Your rights and property will be at peril from new regulations and at the whims of the enforcers.

Hi-Tech Terrorism

The Tokyo subway and sarin nerve gas incident proved that not all terrorists are stupid or without technological training and skills. If nothing else, this brought home to governments around the world that they should not alienate the technically trained people. Japan, as many other nations do, controls the ownership of firearms. However, you cannot control knowledge. In many cases, hi-tech terrorism is made possible by state sponsorship. State sponsorship provides a safe training area and supply base for the hi-tech terrorists.

Other forms of hi-tech terrorism are planting computer viruses, breaking into data banks, and sabotaging data storage archives. Hackers are an archetype of hi-tech terrorism. A good hacker in the National Security Agency computers can make you the Archbishop of Canterbury or a wanted child molester. Even the low-level sabotage of plugging receptors for credit cards can be disruptive. When this is coupled with an ambush (a stationary raid) of the service personnel coming to repair it, it can be really vexing to the authorities.

What is the definition of hi-tech terrorism? The definition is twofold. Most people think of using products of our current technology as high-tech terrorism. This is only partially right. High-tech terrorism also includes using low-tech means to disrupt our hi-tech society. What would these include? At one extreme, it could involve an animal rights group claiming to infect the Thanksgiving turkey with botulism at your local supermarket and at the other end, shooting at the small hornlike devices of the microwave relay systems.

For example, let us look at the perpetrators of the Tokyo subway attack. The Aum cult had sufficient chemicals to manufacture six tons of nerve agents and the ingredients to produce botulism toxin. They had made attempts to obtain the Ebola virus, had tested sarin on sheep in Australia, and were planning a coup in Japan. They also possessed a war chest in excess of $300 million, a Russian helicopter, and two unmanned drones as a delivery system. Their membership was 10,000 in Japan, and 30,000 in Russia, and they had ties with North Korea.

Because our society is so complex, it does not take too many terrorist acts to disrupt it. Major disruptions will have a domino effect. Twenty-first-century technology will be coupled with first-century human reactions. There is also the fear of a terrorist act. This fear by itself will cause a major change in our lifestyles. This may result in poorly attended public meetings, reducing the participation of people in the democratic process. This would enable a small number of activists to hijack the governing of an area.

Day One

When you hear about the hijacking of a French plutonium shipment destined for Japan, a loss of Russian nuclear warheads, or that a group has carried out a biological test on some animals, then you know that Day One is here or very close. For years, we have been hearing about people breaking into the government computers through the telephone. They are called hackers. Until now, this has not been an organized form of terrorism. But that can change.

We already have some evidence of the so-called Class III Information Warfare. Electromagnetic pulses, logic bombs, and HERF (High Energy Radio Frequency) guns enable hi-tech terrorists to bring the computers of banks, financial houses, and trading organizations to a halt. This was already demonstrated in England by successful extortion of at least four banks. What a wonderful way to finance a revolution! On a larger scale, this could be an electronic

Pearl Harbor and may lead to a denial of service by the financial community.

A proliferation of surveillance devices will accompany all government efforts to fight terrorism. However, if hi-tech terrorism breaks out, we will also see all kinds of sensors and warning devices on the market. Entering government buildings, corporate head-quarters, and sensitive installations may require passing through metal detectors, explosive-sniffer devices, and multiple identity checks.

What to do:

- Read some well-developed fiction stories dealing with this kind of terrorism, ones that will provide you with a starting point for what awaits us.
- It may pay handsomely if you invest in the stock of manufac-turers of security services. If you find that a certain type of equipment is sold out and in short supply, that is a very good indication of which stock to invest in.
- Make sure that all your vaccinations are up to date.

Day Two

This is the day when spectacular, but localized incidents grab the headlines. Open warfare between nations may be replaced with hi-tech urban warfare. It is much cheaper and pays just as hand-somely in propaganda headlines. To see if this is the start of a campaign, you must figure out if there is a pattern to the events. The more patterned the attacks are, the more likely that a state-sponsored organization is carrying them out.

What to do:

- If you expect chemical or biological attacks, carry your gas mask with you. If the terrorists are using only one kind of chemical, carry the antidotes with you. Make sure that those with expira-tion dates are still effective. Be like a bachelor, cook according to "best before" dates.

219

Day Three

By this time, it's hitting home that we are dealing with a case of hi-tech terrorism. Naturally, the government will overreact. There may be special permits required to purchase certain chemicals or biological specimens. This permit system will add cost to legitimate users, and you will end up paying for the measures in higher prices.

The hi-tech nature of terrorism will result in additional controls on scientific research and activities. This will result in more and more government controls on technology-related activities.

What to do:
- Obtain all the information on chemical and biological agents used to date.
- If you are dealing with any of the proscribed items, make sure that your paperwork is up to date.
- If you use the Internet, it would pay to have your connection on a second "throwaway" computer. You should also remember that anyone with the know-how can look at all the e-mail you received at your address. Something you wrote in a fit five years ago may come back to haunt you.

Day Four

Because of the paranoia associated with hi-tech terrorism, many innocents will have their doors broken down at four in the morning, while others will be turned in by neighbors with old grudges. The government will give itself many additional powers to contain the threat. The search for the terrorists may seem like the witch hunts of the 1950s. Should you be a target of such attention, your best bet will be to grin and bear the illogical attacks.

What to do:
- Try to find out what the official terrorist profile definition is. If you fit, I suggest a drastic change to your lifestyle.

- If you have had a running feud with one of your neighbors, make peace or move.

Day Five

There may be travel restrictions imposed on the population. In addition, you will find that phone calls may be monitored. A police state is definitely emerging. The authorities may monitor the purchase of certain electronic components. The curtain of secrecy surrounding many hi-tech terrorist incidents will be torn open by spectacular incidents, such as a paralysis of the banking system.

What to do:
- As said before, make sure that all your permits are up to date.
- You may want to have papers as a transport operation. Permits will be easier to obtain for those in the business.

Day Six

The terrorists, in order to continue their activities, will hijack chemical, military, and other shipments. This will result in additional controls on and probably armed guards for those shipments. There will be a proliferation of new police agencies for transport, ports, and the like.

Air travel will take longer, partially because of the preflight security measures. If you take a flight, expect to have your identity verified, your name checked in a computer network, your baggage and body searched before boarding. Roadside checkpoints will be a daily occurrence.

What to do:
- This may be an excellent time to enter the security business. Several high-paying areas will open in maintaining surveillance equipment, sniffer apparatus, and other security devices.
- If you are engaged in anything illegal, avoid obvious chokepoints like tunnels, bridges, or highway intersections.
- If you travel, give yourself extra time for security checks en route.

Day Seven

Having many people suspected of being terrorists or terrorist sympathizers in preventative detention may reduce the terrorist activities. However, after a bout of hi-tech terrorism, no one will feel safe, and government controls will continue indefinitely. A greatly revised terrorist profile will emerge.

We may see scientific research limited to people deemed to be patriotic, so those deemed to be extremists will see their livelihood terminated.

What to do:

- If you fit the new terrorist profile, move and change your name.
- Review books on your bookcases. Any publication that can be construed to have a terrorist use, including old chemistry text-books, should be cached in a safe place.

Day Eight

Society is now run from the top down, all in the name of security. A police state will emerge with all it entails. Freedoms will be curtailed and confiscation of private property will include homes and businesses from those suspected of sympathizing with the terrorists.

How to Survive When You Work for a Target

Do you work for the government, a gynecologist, an embassy, or a religious organization? If so, you may be an unwitting target for terrorists. In this scenario, we must focus on two major issues. The first is safety in the workplace, and the second is how to get there and back. Oh, yes, and then there are the letter bombs. It is

getting to the point where you can't open any mail coming from an unknown source.

The standard advice given to potential targets is to avoid having a routine. This applies to the route you take to work as well as to your time of leaving home and arriving at your workplace. However, you should go beyond that. You have to establish "cutouts." What are cutouts? To give you an example, let me describe the daily activities of an acquaintance working for a target company in an unstable country.

He wakes up in the morning in a house rented by his girlfriend. After breakfast, she drives him in her minivan to an address in town where she backs the minivan into a garage. He gets out of the van and, using the back door, goes through a backyard and enters another backyard. He uses the remote starter for his car, which also opens the garage next door, waits four to five minutes, then unlocks his car door. A young man drives his car out of the garage, and he takes over while the young man drives ahead of him on his motorcycle. At the end of the day, he drives to a restaurant, and, after leaving through the side door, a friend picks him up while the young man takes his car and drives it to his official residence.

Sound like a lot of bother? It is, but he is still unhurt after five years in that country while many of his coworkers have been hurt, killed, kidnapped, or harassed on a daily basis. This is only one example of using cutouts. Use your imagination, and you can dream up a host of others. The poor man's cutout will have to be less elaborate, but it can be done. Let us look at how. There is no set procedure for cutouts. However, we can look upon the major intelligence agencies' use of safe houses, creation of alternate identities, and travel techniques to provide examples of how to go about it.

Day One

Day One comes when terrorist actions are taken against highly visible targets. As news of the terrorist action spreads, there will be copycats and wannabes who may come after you. In most countries, we are currently living in a Day One scenario, while others are already in the Day Two category. As tolerance decreases, violence will increase.

What to do:

- Maintain a low profile. Blend in with the background.
- Prepare a detailed schedule of your daily activities to see if any pattern emerges. Then figure out how to break the pattern. If you are predictable in your activities, you make it a whole lot easier for someone to harm you.
- Using a map, find alternate routes to and from work, and your favorite restaurant (though you should not have one). Drive your alternate routes and note any likely ambush points.
- Get to know who lives in your neighborhood. If you see a kid hanging around when you leave for work, it is possible that you are being cased for a later visit by someone.

Day Two

This is the day when people like you are kidnapped, murdered, and harmed by factions where you live. The intolerance reaches the point of systemic violence. This can occur between antiabortion and pro-choice groups, big business and big unions, or what have you. When the level of violence increases, all those having business with or working for one of these organizations must take precautionary measures. Keep in mind that the police cannot protect you or your family around the clock.

Your employer may offer to you and fellow workers a minibus service to pick you up and take you to work and back home. This supposed security measure will do nothing but provide a larger number of targets to a dedicated terrorist.

What to do:

- If you have family, send them to a safe place.
- Institute your cutouts. Perhaps it is early to do so, but your life may be on the line.
- Drive with your doors locked and the windows up.

- If your vehicle has a parking sticker from an employer that is a target, you may want to remove it and obtain a pass that can be placed on your dash when you park at work.

Day Three

The attacks are becoming organized with pitched battles against security forces. These may even take the form of a guerrilla war. Spectacular hostage situations will arise like the one at the Japanese ambassador's residence in Lima, Peru. If you have to attend public functions, be extra careful. Have extra clothing in your vehicle, or even better, arrive in a small car wearing everyday clothes.

New regulations will be introduced to reduce public gatherings and ban protests. Some streets may be closed to vehicular traffic.

What to do:

- Have ready-to-eat food supplies where you work. Also have fruit juices and soft drinks on hand.
- Have changes of clothing, including shabby evacuation outfits. These items of clothing will help you blend into the general population.

Day Four

About this time, you may decide to sleep where you work, or you may want to work out of your home. Random attacks on expensive cars may become routine during this phase. Some neighborhoods known to be home to target individuals will be selected for arson, drive-by shootings, and other terrorist acts.

Many affluent neighborhoods will become "gated" communities. That is, private security will control entry and patrol these areas. Gated communities will be relatively safe from random acts of violence, but will not be able withstand an organized raid.

What to do:

- Make sure that you have potable water.

227

- Be sure to vary your routine to reflect the changing conditions.
- Do not wear expensive jewelry or other items that may show you are affluent.

Day Five

If the attacks are large scale, you may want to leave the country. This is particularly important if you look different from the local population. The various security agencies will add to the general confusion as there will be turf fights between them. This will further reduce the protection where it is needed.

What to do:

- Have your passport with you. If you need exit or entry visas, have these well in advance. Many consular offices issue visas well in advance if your occupation requires it.
- Have an open airline ticket with you at all times.

Day Six

If the attacks continue, you may want to find another line of work or have a residence compound at work. You must balance the desirability of keeping your job with the risk inherent in keeping it. At a certain point, it's just not worth the hassle. You must be the judge of this.

The attacks may go well beyond targeting individuals who work for targets. It may extend to making hostages of the families of the employees and even their friends. Once this starts, you must revise your safety plans.

What to do:

- Carry your emergency kit with you, and be ready to leave at a moment's notice.
- Make sure that your family is well away from the terrorist activities, and ensure that they are safe from hostage takers. In most cases, the best solution is to move them to a safe area.

Day Seven

By now the situation will either deteriorate to the point of a civil war or the government will have control. If you are faced with a civil war, leave. A polarization between factions has an unfortunate side effect of labeling people. Once you are so identified, you have very little chance of changing the label.

Rat Packs

Today we make a distinction between warfare and crime, but a few centuries ago, the one merged imperceptibly into the other. The distinction between common criminals and the armed forces of an invading state was entirely academic. Rape, murder, and plunder resulted inevitably from a visit by either.

In the Middle Ages, there was no such thing as a local police force. Occasionally, there were hired night watchmen whose job was to stay awake while others slept. But by and large, it was the entire community that undertook the suppression of crime by apprehending criminals whenever they were discovered. The real threat came from outside marauders, and to counter that ever-present danger, there were expandable armies.

Traveling in Italy, you will pass by many tiny towns perched on the tops of hills. These are Italy's famous hill towns, constructed during the Middle Ages to protect civilians from freelance marauders. Tens of thousands of landless cutthroats were brought together by the fourteenth-century wars between England and France, and during the years of truce, unemployed mercenaries roamed Europe forming private armies that plundered and extorted from any community in their paths. The hill towns are monuments to the terror the mercenary armies inspired and to the peasants' determination to protect themselves the best way they could.

Similarly, the French town of Domme was built high on a bluff as a defense against raiders and rat packs. In the days of dirt roads, it would have taken a peasant with an ox cart an hour to go to his work. The Domme was erected in response to Vikings coming up the Dordogne River from Bordeaux. The high bluff gave the residents plenty of warning of approaching miscreants.

Today there are two forms of rat packs. The first is the familiar inner city gang. The second is the outlaw gang preying upon available resources in rural areas. Let us take a quick look at how to identify them through their activities. They come in all flavors. There are the traditional prison gangs, motorcycle gangs, and Crips and Bloods. There are also the newcomers, such as the Chinese triads and tongs, Jamaican posses, Cuban gangs, Vietnamese street gangs, Japanese Yakuza, Colombian drug cartels, and Russian Mafia.

There is a well-established tendency for the young of our species to form themselves into small, tightly organized peer groups or gangs. Once formed, they establish, within the existing social network, territorial rights to a particular area, in much the same manner as adults, since the territorial sense of acquisition is strong in all of us.

Unfortunately, with the increase in the size of metropolitan areas and the accompanying increases of slum and ghetto areas, available gang territory is rather limited. One tangible result of this lack of space has been gang wars, or rumbles, between opposing gangs. These result in very high injury and mortality rates among the participants and even the bystanders. When this different morality of the gangs is intertwined with an inherent viciousness, faster reaction time, and greater resiliency of the youth, not to mention that young people consider themselves bulletproof, you have a real threat.

The rural outlaws are similar to rat packs. The difference is that they are more mobile; bikers are an example of this. Criminal gangs gathered about the armies during World War I and II, Korea, and Vietnam. There are many such groups operating in Third World countries. Somalia and Liberia are two recent examples. They are even more prevalent in areas receiving massive aid from developed countries.

It would be dangerously stupid to ignore the existence of such groups. Warlords may or may not be leading the rat packs. A warlord emerges to provide protection for his area when weak governments cannot control areas of a nation. These warlords, in effect, can set up regional governments.

To illustrate further that even small towns are not safe anymore, all along the rural corridor that parallels Interstate 95 from Florida to New York, the Jamaicans have cornered the crack network. Small Town, U.S.A., offers easy profits to drug dealers at low initial risk because rural communities lack the drug awareness of big cities and are even less prepared than their urban counterparts to cope with naked savagery. Local police forces can be easily overpowered and even more easily corrupted.

To give you an idea of the magnitude of this, take Florida. In Florida, the multibillion-dollar drug business dwarfs all other industries, including agriculture and tourism. In areas like South America, Pakistan, and Afghanistan, where cocaine, marijuana, heroin, and opium are produced, the drug dealers are better armed and equipped than the armies and police forces that are expected to control them.

Day One

Rat packs form due to crises. At first you are not affected. You will read about vicious gang fights breaking out. The death toll will be very high during these gang fights. The recruiting ground for rat packs is among the children of the poor. The day welfare checks are cut back, we will see a vicious trend emerging among the young in the poor sections of the cities. You will not be unduly concerned at first as the gangs are doing in each other, not the general public.

Ironically, self-help law enforcement is more likely to enhance civil liberties than continuing the status quo. Unchecked, drugs will threaten our civil liberties and rights to privacy because they are bringing us to a world where everyone must urinate for everyone else on command! Once we get used to that, additional intrusions into our privacy and civil liberties will be accepted as matters of course.

Compounding this is the problem of police corruption in this country. In the old days, many policemen were on the payroll of gambling establishments. Today, with increased drug profits, we find an even larger police involvement in crime. Drug money finances many a poor country's public services, and in this country, drug money finances larger police department budgets. How so? When you declare war on drugs at a cost of $100 billion a year, the last thing your DEA agent wants is an end to his career by winning that war on drugs.

What to do:

- Be prepared to defend your home! Learn to shoot, and practice to maintain skill with firearms. Learn about the use of cover, the techniques of home defense, and the nature of the threat.
- Be careful in all your dealings with authorities. Your assets can be confiscated under RICO and other laws. All it takes is an allegation that your money came from drug dealing. Complaining to police internal-affairs people is like having the goats guarding the cabbages.
- Place chicken wire on windows exposed to the street. This will keep out Molotov cocktails and homemade pipe bombs.

Day Two

About this time, the government will introduce curfews, and the police will patrol the affected areas in force. The gang fights will change in character. There will be widespread sniping and raiding. The local police will not be able to control the situation and will have to call in the National Guard.

Should you live in an area subject to the raids, your best bet will be to move for the duration. Even better, move permanently to a safer area of the country.

The initial conflict will most likely be in the form of firebombings. This will force gangs to fortify their meeting places and to try to hide them.

What to do:
- Avoid the area of the conflict. Some person may want to see if his sights are on by zeroing in on your head.
- If you live in the affected area, move now. If you have to leave your possessions behind, that is a small price to pay. Sometimes it pays to switch rather than fight.
- Arm yourself in a discreet way.

Day Three

The authorities will try to seize weapons from the gangs. They will also try to register and control all firearms in private hands. As always, this leads to firearms going underground, so instead of knowing who has what, the authorities will know even less. Remember, community law enforcement is the *only* proven method for controlling gang activities.

A black market in firearms will result in burglaries of homes of people known to own firearms. We may even see raids on police stations to obtain firearms.

What to do:
- This is not the time to register your firearms. Cache them.
- Introduce a community patrol program. Use people who are volunteer firemen, Little League coaches, and the like. Do not use thugs. Do not arm the patrols. Use two-way radios to bring help quickly. Obtain some kind of uniforms for the patrol. You should also carry insurance for the community patrol program.

- Provide activities for teenagers. Bored teenagers drift into gang activities. Even better, see if you can use them in the community patrol program. If you do, make sure that adequate training and supervision is provided.

Day Four

The gangs have carved out definite areas. These are no-go zones for the police forces, and even the National Guard will pull out at night. The borders of these gang areas will be marked by burned-out buildings and gutted vehicles. The people living in these areas will be held hostage by the gangs.

The character of the conflict will change from raiding to a defensive posture. The gangs will try to protect their territories. Such a defensive posture will require very large forces to dislodge gangs. The advantage is usually with the side defending.

What to do:

- You can make some money by selling food to these areas, but make sure that you have protection when you deliver.
- Do not associate with any of the gangs. Once you are part of a gang, you lose your freedom of choice.

Day Five

Many people will move out of the gang-controlled areas. The gangs will try to stop this movement. The reason is simple: If only gang members are left behind, then the armed forces will have an easier time identifying them and cleaning up. The armed forces will stage house-to-house searches. These can be very violent. If the armed forces casualties are high, you may see raids by armored vehicles that will fire at any sign of life. As usual, the innocent will comprise the majority of the casualties.

Anyone admitted to a hospital with a gunshot wound will have some explaining to do. The number of detention camps will increase as the violence continues.

What to do:

- If you decide to stay in a contested area, form a protection force and patrol your area. The protection force *must* have an armed back-up unit available to react quickly to any gang incursion into your territory.
- Maintain a communications watch to keep track of what is happening in your area.

Day Six

The police will pull out of gang territories. This will encourage the gangs to set up territorial, warlord-type governments. The warlords may or may not provide protection for the people living in their areas. The government, in desperation, may decide to use area weapons against gang-controlled zones. If this happens, even more innocent lives will be lost.

What to do:

- If your protection force is not adequate to the task, evacuate the area.
- Do not form alliances with gangs. Your protection force may end up as cannon fodder for a turf dispute with another gang.

Day Seven

The gangs will set up their own temporary governments in the areas they control. The government, in response, will mobilize the reserves to combat the rat packs. These engagements will bring the use of armored vehicles in gang-controlled locations. The gangs can defend only urban areas, where armored forces are working at a disadvantage. The rural gangs will have been eliminated by this time. Dispersed, they may make for the cities or wait until the army pulls out.

What to do:

- If your group lives in a contested area, this is about your last chance to leave the area safely. This must be done by the whole group. You may have to fight your way out.
- Try to determine which areas are likely to be free of gang activities in the future. You and your group may want to move to such an area.

Day Eight

The government will occupy the gang areas with regular troops. The gangs will disperse and take up residence in other areas, starting the cycle all over again. Since the government has limited resources, the gangs may trickle back to their former territories.

What to do:

- The newly cleared area may provide you with improved living accommodations. Evaluate and move, if feasible.
- You may make money by rebuilding burned-out properties.

Day Nine

As a reaction to the gang wars, the government will enact new laws and regulations. These will limit the number of people congregating at a place. (Religious worship may be exempt, in which case we shall see the formation of ".44 Magnums for Christ" congregations.)

It is interesting to note that in the sixteenth century in the Middle East, in what was later to become Israel, Jordan, Lebanon, and Syria, there was very little government. Having little government, they had no major wars. Contrast that with today. War everywhere! The breakup of the Soviet Union started a series of small-scale wars. Many new countries came into existence, appropriately called Chaostans. Now there are some 500 nuclear arms unaccounted for. That is more than enough to arm every terrorist group, nationalist faction, and two-bit dictator, just as ethnic rivalries are flaring up in Europe and all over the world.

More than 100 new nations have been born in the last fifty years, with race and religion shaping the future as never before. The threat of Total War has diminished now that Russia and the U.S. are "friends," but the horrors of smaller wars are now much worse, and the threat of nuclear terrorism grows daily.

While some of the wars going on now are familiar to anyone who follows the news, there are many tribal and ethnic conflicts you probably haven't even heard of yet. There were thirty-three countries at war at the beginning of 1997—covering nearly half the world. That doesn't even include U.S. involvement in Bosnia or Haiti or many of the military interventions now going on. Why is this so? Because large, centrally controlled organizations that used to keep things quiet—like the Soviet Union—cannot control or compete in

5

Warfare Survival

the new world. Soon there will be 100 different wars being fought across the globe, and this doesn't include internal small-scale violence in many Third World and newly independent countries.

There is an opportunity in all this and a threat as well. Many of the combatants will view the U.S. as the friend of their enemies and will try to strike back either through military or terrorist actions. Just think of the Iranian ayatollahs calling the U.S. the "Great Satan."

What shape will this conflict take? The use of conventional weapons is most likely, first because they are available cheaply and in large numbers, and secondly because of the fear of superpower retaliation if NBC (nuclear, biological, or chemical) agents are used. However, the losing side will resort to the use of whatever is available to them, probably conventional attacks on the enemy's nuclear power generation facilities.

We see on one hand a worldwide effort to reduce arms held by the population of most countries, but on the other, arms are readily available for any government or organization. This illustrates the hypocrisy of the arms-producing countries. What they are saying is that it is quite all right to have government-sanctioned killing, but people defending themselves is discouraged.

In a warfare scenario, dictatorships and other highly centralized societies have an advantage over democracies in that they are able to make decisions more quickly. Hitler's initial successes are, in part, due to this. However, as the war went on, the few months' advantage in decision making lost its importance, and the rest is history. In today's missile age, C^3I systems are supposed to compensate for this. Let's hope we will never find out if they work. But what if the situation changes and the decision to double the number of cruise missiles can become a reality in a few days? Then those few days become critical.

The potential for invisible weapons is emerging with miniaturization. This trend will spawn more and deadlier wars than in the past. The "nanosystems" (nano equals dwarf) described by K. Eric Drexler, the founder of the Nanotechnology Study Group at MIT,

will change the face of the globe forever. Some of the technology is already out there. However, just possessing the technology does not mean that a nation will use it to its full advantage. Revolutionary change like this has happened before, and the first power to use it has won and won big.

For example, take the Industrial Revolution. In one century, Britain went from an agricultural to an industrial society. The urbanized Britain became Great Britain as its head start enabled it to dominate global politics and forge an empire that lasted more than 100 years.

Today a computer chip is nothing but silicon dioxide with impurities added. The minute quantities and shapes of the impurities define what a computer chip can do. Yet the "dirt" must be added in absolutely clean rooms. The advent of the computer brought with it the age of miniaturization. Small computers replaced bulky control systems in all devices we use today. With smaller size came increased "bang for the buck" in warfare, and today we are working on weapons that are almost undetectable. This leads to a newly destabilizing world.

To illustrate the destabilizing trend, we need to look no further than India and China. Both are huge countries that have fought each other in the past, both have nuclear weapons. But let us look at where they are headed in terms of using emerging technologies. China is a closed, monolithic society with emerging markets and a very primitive infrastructure. Chinese students studying abroad tend to remain in the countries where they study. In contrast, India has English for a second language and Indian students largely return

to their country. Indian technology is innovating at a fast pace, and Indians are converting their country into a modern state. India also has a record for entrepreneurship. Look no further than communities in the U.S. with a significant Indian population to see the number of small businesses they own.

Miniaturization will reduce reliance on fossil fuels. Who needs supercarriers and their aircraft when a patrol boat could deliver devastation equal to that of a carrier? Should this happen, oil demand would fall, and the Middle East would lose a great deal of political and economic power. Such a technology will not be embraced by states like Libya, Iran, and Iraq. These principal terrorist states would go from the center stage of the world to near oblivion.

Since they are nearly invisible, these weapons could truly be programmed to be genocide weapons. For example, if released from all sides of the border, they could be made to kill every living thing they encounter, thus wiping out an entire nation. Being nearly invisible, they can't be the subject of disarmament agreements. How can you ban something if you can't verify its existence?

This near-invisibility will destabilize the world. Perhaps in time our ability to detect the new weapons will increase, but generally it is easier to hide small things than to find them. On top of these problems, these weapons will more than likely be manufactured in small plants. The enemy will have no smokestacks to aim at. Thus, of the three traditional targets—factories, weapons, and people—only the people will be visible.

Enough of the future. In the near term, we will have to cope with weapons on the shelf. That is what this chapter is all about.

Day One

Given the daily fare in the media about wars in one place or another, we must distinguish between purely local conflicts and those that may spill over to this continent. Those likely to spill over to America are characterized by:

- Use of nuclear weapons
- Extensive naval activity to disrupt shipping
- Use of chemical and biological weapons
- Blaming the U.S. as a supplier for one of the participants

If the war is with Canada or Mexico and you live close to the border, you may want to vacate your home for the duration. Neither

country is a significant military power, but this does not preclude a local conflict resulting in loss of life.

What to do:

- Have your emergency kit ready and stocked.
- Cache firearms, ammunition, information on various subjects, food, equipment, water filters, and the like. You may need these later on.
- It is a good idea to store seeds for your "Victory Garden."
- Certain foods, goods, and metals will be diverted to the war effort. Try to obtain these by scavenging, recycling, and other means. You can resell them to private industry.
- Be sure that you have equipment to keep up on international news. A multiband radio is best for this purpose. The Internet is questionable. With a little effort, the Internet can be monitored and participants identified. After all, it was designed as a safe military network.

Day Two

By this time, you have an idea on how the war is going and what kind of war we are in. If you hear that a mutual disarmament treaty has been concluded and the opposition is sending "observers" to America, you know that we have lost. Act accordingly.

How to Cope if We Lose

The United States is the most powerful nation on the face of the globe. Even in the time of nuclear weapons, the ultimate goal of any war is to occupy the enemy's territory. What would happen if we lose? There are many movies depicting the German occupation of European countries. Thankfully those situations were temporary. After America entered the war, everyone knew that eventually the

Germans must withdraw. However, should the U.S. be occupied, being the last of the superpowers, we can look forward only to a long-lasting domination of the country. It really does not matter if the occupiers are Chinese or aliens with little green beards, the effect is the same—their culture will be dominant. How can you cope?

As always, there will be some who cooperate with the occupiers. When the Allied troops occupied Germany, many women openly propositioned soldiers to become their lovers. Why? First, by having only one man to satisfy, they could avoid repeated gang rapes, and second, they had a food provider. If this could happen there, it will happen here. Then, of course, there are the faint souls who will give up at the first sign of a cloud in the sky. Given the possibility of collaborators, one must form resistance cells and fight against the occupiers in a circumspect way.

Our traditional defense, having an ocean on either side, has protected us in the past. What the future holds is not known. If an enemy occupies Mexico or Canada, we may have to face the specter of a land invasion for which we are poorly prepared.

Do not forget that to the occupier's troops, America will appear like a dream world. We do have one of the highest standards of living in the world.

Day One

This is a hard one to call. In general, when a nation's elite hold their sons out of the military service, or when foreign troops are hired, it shows that nation's loss of will to be free. A government not trusting its population to be armed is another indication of a deep rot in that nation's morale. These are the initial signs of a doomed empire.

Another possible start to Day One is a greatly enhanced United Nations. If the UN had its own army, we could see UN troops trying to ensure that U.S. wheat is sent to starving Third World nations.

What to do:

- Follow current events closely.
- Lay in plenty of contraceptive pills for all females of child-bearing age in your family.
- Start to prepare hiding places and cutouts as explained earlier.
- Purchase at least one battery-operated shortwave radio receiver. Be sure that you have plenty of batteries for it.
- Lay in supplies of pantyhose, soap, razor blades, batteries (lithium are best), sugar, salt, flour, matches, tea, coffee, needles and thread, toothpaste, candles, powdered milk, dehydrated soups, and, most importantly, vitamins. These are your trade goods and supplies.
- Purchase a good, sturdy bicycle for each member of the family, and have spare parts, tires, and a repair manual. You will need these later.
- Have some of your firearms, spare parts for them, ammunition, cleaning supplies, and a portable reloading unit in a safe cache.

Day Two

The occupiers arrive first in the large cities and military bases. Major national figures, such as members of the House and the Senate, will be sent to "reeducation camps." Oil imports will be discontinued, and fuel ration cards will be issued for essential travel. The issuance of new identity documents will begin with an interrogation by the occupying forces. Random shootings, either of hostages or as a result of mistaken identity, will be daily occurrences.

The occupiers do not want to kill all Americans. They just want a submissive population to deal with. There will be major battles with those who would much rather fight than have their lives snuffed out in some dark cellar of the new secret police. This hopeless resistance will not be of value to the country as a whole.

America, having the highest per-capita private ownership of firearms in the world, will resist. Therefore, the occupier will use its armored units to seize gun shops and will use the BATF forms to track down individual gun owners. This will be a backward process; that is, they will start with the most recent purchases and work backward. A dusk-to-dawn curfew will be declared. The use of classroom informers will be prevalent.

What to do:

- You must decide during this phase whether to stay put or go abroad. The occupiers' control will be loose enough to cross the borders with ease at this stage.

- If you belong to any patriotic organization, are a former member of the armed forces, or are in a position of high visibility move and move fast. Acquire another identity.

- Keep women off the streets, and have them avoid being seen through windows. Given past incidents in history, massive rapes will occur. Usually six or more soldiers will enter a house, and while the men are held at gunpoint, the women are raped by the soldiers. If the men fight back, the soldiers will kill them, and the women will have no providers.

- Have a back way out. If you hear the front door being broken down, try to escape out the back. Have emergency hiding places prepared for all members of the family.

- Wear your oldest, shabbiest clothes. Do not wear jewelry or a watch.

- Hide all your possessions as well as you can.

- If there is looting or rape, do not complain to the occupying authorities. They will classify you as a troublemaker and later will focus the attention of the secret police on you.

- Learn the occupiers' language. This can be one of the most important steps to take.

- When you see occupier troops while walking on the street, step aside or step off the curb, avoiding eye contact. Do not

attempt any kind of heroics. There will be better chances later on to get even.

- Do not discuss anything in front of your children. They will be questioned in school, and anything they might say will be reported later to the occupying forces.

- Always have a small bag with warm clothes, boots, socks, and some high-energy, long-lasting food ready. This is in case the occupier comes to pick you up and send you to a concentration camp, or worse.

- Have all your firearms and related items hidden. If you are known to be a hunter or sportsman, immediately relocate to where you are not known. Someone in your old neighborhood will surely turn you in sooner or later.

Day Three

The occupiers will spread out into the smaller towns. At this point, certain people will be picked up and deported to the occupiers' homeland for forced labor. The most likely classes to be deported are former government officials, members of the armed forces, the judiciary, business owners, and officials of major corporations. Major production and engineering equipment will be dismantled and transported to the occupiers' homeland with the skilled technicians *attached*.

The occupier will, as in past occupations in other lands, pick a minority from which to recruit the local government employees, policemen, and other functionaries. There will be a new "democratic" government. Only later will complete occupier-control come. The occupiers' control is very thin at this point.

The major economic activity now will be the setting up of labor camps, which will happen big time. There will be arrests of "war criminals." History has proven over and over that the victors write the history books. Yesterday's general becomes tomorrow's warmonger.

The confiscation of homes will start. The occupiers will first take luxury homes and then billet people in more modest homes.

THE SURVIVALIST'S HANDBOOK

House-to-house searches will be made to confiscate firearms, gold, jewelry, and anything of value.

What to do:

- During the first few periods of the occupation, besides the random looting, rape, and murder, the occupiers will have no time for the small stores. So buy everything in sight, laying in as many barter goods as possible.

- Try to get as many of the new occupier identity documents as you can. This will also give you additional ration books. Purges of the antioccupier elements will continue. By reading this book, you are already one of them.

- Take stock of what is happening, and change your occupation, if necessary, to prepare for Day Four. It may be advisable to become a farmer.

- Before someone is billeted with you, have trusted friends or family members move in with you.

- Make an effort to make your home look shabby and run down. This is not the time to paint your house bright green or pink. That may invite unwelcome attention and visitors.

- Buy all the medical supplies and drugs you can lay your hands on.

Day Four

Now it is the countryside's turn to welcome the "liberators." Major firms will be nationalized. Savings will be wiped out by the "currency reform" the occupiers will institute. American grain and other food-stuffs will be sent abroad in great quantities.

At this point, the safe occupations are: accountants, artists (tame), athletes, barbers, civil servants (low level), dentists, doctors, dressmakers, farmers, firemen, industrial workers, librarians, local policemen, musicians, optometrists, photographers, printers, and schoolteachers. These will be excellent times for psychopaths, poison-pen writers, sadists, the new left, socialists, and quislings.

Occupations that are likely to have you "volunteering" to work in the occupiers' country are aircraft workers, computer programmers, engineers, scientists, and technicians. If you were an academic, businessperson, ecologist, economist, journalist, lawyer, military personnel, psychiatrist, trade unionist, or a veterinarian (who can afford pets now?), your job prospects will be bleak. Similarly, the anarchist, Chinese-American, clergyman, drug addict, Japanese-American, Jew, and member of the John Birch Society will fare very badly.

The occupier will, at leisure, look for contributors to conservative causes, holders of explosives permits, current or past federal firearms license holders, anyone with past federal police experience, and owners of registered firearms. Once found, these people will also swell the numbers of the deported. Those who belonged to any patriotic organization like the Boy Scouts, Girl Guides, Shriners, Lions Club, or more particularly, veterans organizations will be identified for later detention.

People whose parents emigrated from the occupiers' country will be used initially as translators, but will be suspect and eventually will be sent to labor camps.

What to do:

- By this time, you should have a barter network functioning.
- Start forming small sabotage units. Do not leave empty brass around for the forensic people to trace you.
- You really should be out in the country by now.
- Keep an eye on people moving about during the curfew. Unless they are in an essential occupation, they are either informants or resisters.
- Learn all you can about substitutes for imported food, such as using chicory for coffee or making maple sugar.

Day Five

New agricultural policies will be introduced. The occupier may even use farm collectives for large-scale production of foodstuffs. The food may be used by the occupier or sold for hard currency. You are not likely to see too much of what you produce.

The identity cards will be replaced by internal passports. These will be issued after rigorous checks and interrogation. The document will record changes in address and employment, which must be approved by the local police. Border exclusion zones will be established. These are usually ten- to fifteen-mile strips. Special permits will be issued to those living in these zones. To visit these zones, you will need a special permit, issued only after exhaustive checks.

The use of informers and police agents will be widespread by this time. You will be judged guilty until proven innocent. Confiscation of all radio equipment will be almost complete. The occupiers will monitor citizens, band radios and walkie-talkies. Use of such equipment will be evidence of resistance activity.

What to do:
- Your additional identity documents will become more or less useless unless you have built a strong cover for their issue in the first place.
- Have contacts in the countryside to obtain fresh produce and other foodstuffs.
- Collect old newspapers, which can be rolled up to make substitute fire logs.

Day Six

By this time, the American economy will be thoroughly disrupted. Most businesses producing anything other than the barest necessities will collapse. On top of this, all vehicles will be restricted to official purposes only. The use of a personal car will be for state-sanctioned carpools only.

If you hear that some officials in the Department of Agriculture have been convicted for anything, then it is safe to assume that a famine is coming. The internal passports now will require travel permits to visit other states. The newly beefed-up transport division of the internal police will routinely ride trains and have checkpoints along major roads to ensure that your permits are up to date.

The occupiers will have quick-reaction forces to deal with insurgents and sabotage. These forces will commit atrocities, thus adding to the misery of Americans. These forces will be modeled along the lines of the KGB internal and border units.

What to do:

- With any luck, you can choose like-minded carpool members. Learn how to tamper with the vehicle's odometer and to obtain additional gasoline.
- Start to snipe at individual collaborators, but do not undertake large-scale raids. Even better, sabotage their cars.

Day Seven

About 4 percent of the remaining population will cooperate totally with the occupiers, giving them several million reliable people to run America. This, of course, will be under the direction of "advisers" from the occupying power. This period will see the start of organized resistance movements and perhaps even some liberated areas. By now all private cars will be under state control, except those used by farmers and deemed essential. This is not a big deal since gasoline will not be available to private individuals. The "grounded" population will be easier to control. Checkpoints leading into towns and cities will be permanent.

If the occupiers maintain computer links, the use of forged documents will only be a "pass directly to interrogation." The computer links can be sabotaged at will.

What to do:

- Start to use your bicycle. If it's a flashy model, spray dull paint over it to make it appear shabby.
- Plant a garden. Try to camouflage your garden by planting corn or something similar around it. Your garden will attract midnight looters as food supplies diminish.

- If you have any gasoline left, hide it for later use. You will need gasoline for gelled fuels and Molotov cocktails.

Day Eight

Most likely there will be an immense expansion in the number of economic and administrative bureaucrats, recruited primarily from the chosen minority. There will be full employment for them, at the cost of efficiency. The raw materials allocated to your firm may have been wrongly allocated. They will never arrive on time, if they arrive at all.

What to do:
- If you are still living in a city, try to move to the country.

Day Nine

American "volunteer" troops will be sent abroad to fight "liberation" wars for the occupiers. Given that, during the purges, more men have died than women and that "volunteer" labor and military units are mostly comprised of men, women will operate most of the farms and supply the industrial plants. Small stores will be nationalized during this period.

What to do:
- If you are a woman, it helps to have a good trade or profession. Otherwise prepare for hard labor.
- Small armed groups should visit the bureaucrats at midnight and point out that they can regulate as much as they want, but that they should not enforce those regulations at the cost of their lives.
- Having occasional accidental deaths among the occupiers' American functionaries will ensure lax enforcement of the regulations.

Day Ten

A modern-day feudal system is emerging. There is the fighting elite (the occupiers), the priests (the turncoats), and the serfs (the population of America). The serfs provide the means for all three classes to live. In the meantime, the propaganda machinery tells the serfs how much better off they are. If you had been in the so-called capitalist class, there will be almost no chance for your children to obtain adequate education.

What to do:

- Working out of your basement workshop to repair or build appliances, formulate chemicals, or even to teach children will give you additional resources to provide for your family and friends.
- Make the occupiers stay in their garrison. This can be done by ambushing small units.

Day Eleven

Once the occupiers have to patrol in force, their control over the countryside will loosen. This will permit the training of more people for guerrilla groups. Some areas will be free of any land forces at this point, most likely those not involved in large-scale food production or manufacturing activities.

What to do:

- Prepare a warning network in your neighborhood. This should alert you to any occupier incursions in your area. You can use low-power walkie-talkies or CB radios. Depending on the size of the units, you can decide whether to fight or hide.

Day Twelve

It's time to throw off the occupiers' yoke. Conduct guerrilla warfare in urban and rural areas. Harass the enemy forces, sabotage their warehouses and communications facilities, and start sniping raids into their headquarters areas.

What to do:

- Working only with people you trust, carry out large-scale sabotage operations.
- Keep the occupiers in a state of high alert. They can maintain this only for a limited time, and as a result, fatigue will set in rapidly. This will make them easier to subdue.

  The American people must liberate themselves, for now there is no other superpower left to come to our rescue.

Nuclear Warfare Survival

Although, with the meltdown of the U.S.S.R., the possibility of Total War has lessened, the chances of regional nuclear confrontations and the possibility of nuclear terrorism have increased tremendously. As long as there were two superpowers, even though they threatened and pointed missiles at each other, they achieved what military forces were supposed to do. They deterred attack—until now. But, deterrence only works when the other side is convinced that we are both willing and able to inflict harm. This is why we got dragged into the Korean conflict. They did not believe us. Spy satellites are a blessing in disguise, for they let the enemy see that we are strong.

The scenario for a nuclear confrontation can come about very quickly when an opponent having nuclear weapons realizes that in a conventional confrontation, it would lose. Then it may resort to firing all its nuclear weapons at the U.S. mainland. This would be done to destroy and disrupt U.S. troop movements, embarkation of

troops and equipment, and to knock out the U.S. system of deterrence. The escalation ladder prepared by Herman Kahn, one of the foremost thinkers on the American defense system in the 1960s, would be of no use in this case. In a nuclear war, the Ambrose Bierce comment, "War is not fought to determine who is right, but who is left," is especially true.

The arrival of nuclear weapons has given mankind the ability to destroy the planet's ecosystem. Fallout is not the culprit. It is the possibility of the destruction of the ozone layer. Without the ozone layer, life as we know it would not be possible. This, of course, is only possible as a result of a major nuclear exchange. A war between Russia and the U.S. could bring this about. Smaller nuclear exchanges, like one between India and Pakistan or a sabotage of a nuclear power plant, would have much less effect on the U.S. mainland.

There is one subject that merits serious consideration, EMP (Electromagnetic Pulse). In the early days of nuclear weapons testing, this was mostly ignored. EMP is produced in the first milliseconds of a nuclear explosion. It affects most electronic devices, fries transistors, and disables almost all communication devices and computers. Many of us are surfing the Internet and have many of our reference books in the form of CD-ROMs. The Internet will be history and all of your reference books on CD-ROM will be unreadable. If you want information on how to rebuild civilization, keep your books on microfiche or in old-fashioned book form!

Almost everyone exposed to less than 200 roentgens (a measure of the quantity of radiation in the air) will recover. If subsequent daily doses are in the order of six roentgens per day, no incapacitating symptoms should result, although later in life, cancer and leukemia rates will rise. Given a nuclear warfare scenario, immediate survival by reducing the radiation you are exposed to is paramount. Down-range effects of increased radiation will pale in comparison to the immediate effects of a nuclear conflict.

Nuclear reactor accidents have some similarities to nuclear war. First there is fallout, which does not follow the normal decay scenario. Fallout from a nuclear reactor incident is much more persistent than one from a nuclear explosion, so you must evacuate the area subject to high radiation levels. In a nuclear war, the radiation levels from fallout will be one hundredth after two days and one thousandth after two weeks. Fallout from a nuclear reactor accident could force you to wait months in your shelter. Your best bet is evacuation to a safer area. This is where the availability of radiation-measuring instruments really pays off.

There is always the possibility of a poor man's nuclear war. In a way nuclear terrorism represents a greater threat to our society than does traditional military action. A nuclear device can be assembled in a motel room or an apartment. A crude nuclear device once floated in on a barge, and today Iran is acquiring intermediate-range ballistic missiles from Russian underground markets.

Day One

When you hear about prolonged famine or drought in a potential enemy's lands, when it withdraws its embassy personnel, or when a major conventional conflict erupts between the U.S. and a potential enemy, then Day One may be near. Another possibility is that two other powers at war with each other might lob a few missiles at the U.S. just to spread the misery around.

Watch for other possible scenarios triggering an alert that a possible nuclear confrontation is likely. When you learn that the U.S. is threatening a preemptive strike against North Korea, it is an open invitation for a poor man's nuclear war on America. They could do it, too. Both Mexico and Cuba have large numbers of North Korean fishermen, advisers, and diplomats. There are also the perennial nuclear wannabe powers of Libya, Iraq, and Iran. Add to them the possibility of Hamas employing some Russian expatriate nuclear physicists.

Once you decide that a nuclear confrontation is possible, you must decide whether to bug out or batten down the hatches and ride it out. Staying where you are has tremendous advantages. You

have access to your supplies, shelter from the elements, and you can protect your stuff from looters.

What to do:

- Decide whether to evacuate or stay put. If you planned ahead, you should be living in a rural area by now.
- Use your time to prepare a fallout shelter.
- Make sure that you have a radiation rate meter (a digital one is best), a dosimeter, and chargers and batteries to keep the equipment functioning. Most chargers operate on a single battery, or there are those that are completely manual.
- Buy more food and trading goods.
- Make sure that you have heavy-duty wrecking equipment to get you out of your shelter in case the entrance is blocked.
- Having a vehicle from the sixties or seventies, which will have points instead of an electronic system, can keep you going. EMP will fry most existing electronic systems dependent on transistors.
- Have a library of how-to, technical, medical, and other publications in microfiche format. Store these in old ammunition cans.
- Lay in supplies of potassium iodide and L-Cysteine. Potassium iodide prevents the absorption of radioactive iodine from fallout by the thyroid and should be taken for sixty days after a nuclear detonation. L-Cysteine provides some protection against radiation and should be taken as soon as news of potential attack is received. Take 130 mg of potassium iodide a day. One thousand mg of L-Cysteine should be taken the first two days and 500 mg per day thereafter.
- If you have any warning, remove all loose objects, such as potted plants, from your yard.

Day Two

If there is any warning, this is the time to evacuate the target areas. There will be mayhem on the roads. If you can, evacuate by boat. If not, find shelter, or if you have one, move your additional supplies into it.

Do not join any mass evacuations. Most small towns will barricade themselves, and the evacuees from the city will turn into a mob. Kurt Saxon, one of the first authors on survivalism, called them "Killer Caravans." In such a case, you and your family will be stripped of your hard-earned supplies, and you will be defenseless.

A few more words on electromagnetic pulse. During the detonation of a nuclear weapon, EMP is a radiated electromagnetic broadband pulse with a very high energy content. If this occurs at a high elevation, the interaction of the weapon's radiation with the upper atmosphere (partial vacuum) will cause a larger area of radiation and propagation of this pulse than if it were a surface detonation. A surface event will have a faster rise time pulse, but will affect a more local area due to the earth's curvature.

The radiated radio frequency pulse will be intercepted by any and all conductors and transformed into a voltage. Power lines, telephone lines, and antennae will capture enough energy to cause equipment failure. Protection of your equipment is a must from both EMP and lightning. You can use similar types of protectors. The difference is in the rise time of the events. EMP is in the low nanosecond range while lightning, at its fastest, is at least ten times slower. A building can handle 100,000 volts while electronics will be damaged with just a few volts.

What to do:

- Disconnect radios and other communication devices from the power supply and their antennae. You can place them in a makeshift Faraday box, an old freezer wrapped in aluminum foil.

- Fill up containers with water. Line bathtubs with a clean shower curtain or caulk the drain with silicone caulking. The liner serves a double purpose. First, it protects the water from contamination from the bathtub. And second, it seals in the water, as most bathtub plugs leak.

- Move supplies to your fallout shelter.
- Add additional layers of protection to your shelter. If it is in the basement, cover basement windows with plywood and shovel earth over them.
- Make sure that you have batteries for all electrical equipment.
- Disconnect water, gas, and electricity. Make sure you obtain the tools to do this and that all family members know what to do.

Day Three

Now you know that the attack is underway. This comes by a warning from the authorities or by visual observation. If there was a mass evacuation, the mobs will have stripped bare the countryside within a tankful of gas drive from the city. This will result in the farmers being hurt or killed, thereby endangering future food supplies. Urban people are not noted for their farming skills. Survival during this period will be from stored supplies. It is now too late to go shopping.

The most effective way to reduce losses of health and life from radiation sickness is to prevent excessive exposure to radiation. *Adequate shelter and essential life-support items are the best means of saving lives in a nuclear war.* Do not confuse the emotional reactions in a disaster situation with radiation sickness. The best way to avoid this is having a personal dosimeter. Nausea, vomiting, headache, dizziness, and general feelings of illness can be reactions to stress as well as the first signs of radiation poisoning. In addition, in both cases these symptoms end within a day or two. Diarrhea from common causes may be confused with the onset of radiation sickness, but hemorrhages and substantial loss of hair are clear indications of having received serious, but not necessarily fatal, radiation exposure.

What to do:

- Keep a log of radiation levels in your shelter and outside of it. Use a remote probe for outside measuring, if you have one.

(If you do not have a remote probe, do not go outside for any reason.) Take a reading every ten minutes.

- Start your filter system. Remember that you need five cubic feet of air per minute.
- Treatment of people exposed to radiation currently consists of the following:
    - Blood and platelet transfusions
    - Use of antibiotics to prevent infection
    - Use of plasma and electrolytes to maintain fluid balance and blood volume
    - Sedatives, tranquillers, and painkillers to control anxiety, pain, and to get rest
    - Give bone marrow suspensions or bone marrow transplants to fight low white-blood cell count
    - Antiemetics to control nausea
    - Maintain a good diet with vitamins to support the above
- If you have a radio with a short aerial, and if this is backed up by another in safe storage, you should monitor reports on what is happening in the nation and on the international scene.

Day Four

This is the time to wait it out until the radiation levels drop. You must wait until one hour outside will expose you to no more than five rads (the amount you absorb) of radiation. The recommended exposure limit of five rads per year will be passé. Your best bet is to try to keep your accumulated radiation exposure to less than fifty rads. Even at this rate, in later years leukemia and other radiation-induced diseases will be higher. Almost all of those receiving a 600r dose will die within a few weeks.

The radio spectrum may be very poor due to the crap in the atmosphere. EMP, being a very short-lived event, should not be a problem unless there are follow-up nuclear attacks.

What to do:

- Make sure radioactive particles do not block air filters.

- It is a good idea to be able to switch filters without having to go outside. Be sure the used filter is not a source of radiation inside your shelter.
- You will see the first effects of what is called "survival guilt." Many survivors will be apathetic and wish they were dead instead of their loved ones. You must keep busy and keep your group busy to reduce the effects of this psychological crisis.
- Take stock of your supplies, and lay out a rationing plan for those items in short supply.

Day Five

Now you will go outside to assess the damage to your home and to erect long wire aerials to listen to radio broadcasts to learn what shape the nation is in. Make sure your filter intake is washed down, or at least wiped off, to remove fallout particles. Be very careful not to bring radioactive particles into your shelter. It is best to leave clothing outside.

It is time to take inventory of your supplies. Make lists of what you will need to survive in style. Any item that might have been exposed should be disinfected. This can be done by washing with a detergent solution and then using your radiation meter to check that it contains no radioactive particles. This verification can be done only in your shelter, because background radiation outside would mask the presence of any hot spots.

What to do:
- Take a reading every hour and keep marking the results in your logbook.
- If the radiation is above five rads, do not stay outside for more than ten to fifteen minutes.
- If your home is intact, hosing down your roof can remove radioactive particles. Using a hoe, you can remove contaminated surface particles and move them away from your shelter. Be sure that you remove your clothing before entering your shelter. Decontaminate any hot spots on your body.

Day Six

Find out what has happened in your neighborhood and assess the damage. Determine if surrounding areas have less residual fallout by taking a survey of radiation levels. If you've been listening to your shortwave radio, you should have a good idea whether it was a limited or a general nuclear conflict.

Talk to other survivors, note their condition, and find out which areas they have been through and the condition of those areas. This information will help you form a picture of what is happening around you. Listening to citizens band and other short-range broadcasts will help you to formulate plans for the future.

What to do:

- Survey your area with a radiation meter, and plot the readings on a topographical map. Repeat this daily. The results may suggest relocating to another area.
- Take 130 mg of potassium iodide and 500 mg of L-Cysteine every day. Saturating your thyroid with excess iodine will reduce the likelihood of accumulating radioactive iodine in the thyroid.

Day Seven

If the nuclear confrontation was a limited war, slowly the country will get back to a semblance of normality, but it will not be the normality we take for granted today. To give you an idea what to expect, a study in 1962 looked at casualties from a nuclear war and the recuperation period required. To interpret the table, at its simplest, if we lose 100 million people, our technology and standard of living will be that of the 1890s at best, and if the EMP damage is large, we will be back to Morse code on telegraph wires for our communications network, CB radios notwithstanding.

Now all this is very theoretical, and it was done more than thirty years ago. Furthermore, the country was much less urbanized in those days and as such much less dependent on transportation networks for food. We may see a much higher death toll than antici-pated in the study because of secondary causes such as starvation and exposure to the elements.

Dead	Economic recuperation period
2,000,000	1 year
5,000,000	2 years
10,000,000	5 years
20,000,000	10 years
40,000,000	20 years
80,000,000	50 years
100,000,000	100 years

What to do:
- Start to forage for supplies and tools to rebuild.
- Add other survivors to your group. Accept only those individuals who have something to contribute.
- Update your maps as you gain information. Mark destroyed bridges and overpasses on your local maps and target areas on the national map.
- If you are hunting, discard bones and bone marrow, meat close to the bones, thyroid glands, livers, and kidneys. From plant sources, wash fruits (peel if possible). Tubers must be thoroughly washed. This way you can reduce the intake of radioactive substances. Do not hunt sick animals!

Day Eight

More than likely the federal government will exist in name only. There will be a number of regional governments running the country. It may even fall to individual municipalities to issue regulations and their own currencies. Local warlords will emerge. The efficient ones will not be politicians—the local police chief is a likely candidate.

What to do:

- Time to form a community of like-minded people to grow food and to protect your homes.
- You must have some kind of protection force to ensure that the fruits of your labor are not taken away from you.

Day Nine

The future unfolds, and given the nature of nuclear war, it will not be a pleasant time. Many occupations will be of no use, while at the same time new occupations will emerge. Many of these will be in the area of nuclear decontamination and radiation medicine.

What to do:

- Prepare detailed maps of residual contamination. This will enable you to plan your agricultural efforts.
- Continue to measure residual radiation, as many areas will become safe as time goes on and the radiation decays.

Day Ten

A new society will emerge and the long road to recovering lost knowledge will begin. Trade will begin once again. Trade is a win-win situation, one of the few that unite the world.

Chemical Warfare Survival

During Desert Storm, the population of Israel feared that chemical-weapon-tipped SCUD missiles would once again unleash what Israel fears most, toxic chemicals. Israel, of all nations, because of the Holocaust experience, fears and abhors chemical weapons. The SCUDS were not armed with chemical payloads or did not work, but it was a close call.

Chemical warfare can be government sponsored or freelance, that is to say, akin to the Tokyo subway scenario. The outcome is the same—civilians dead or maimed in the name of an obscure cause. Certain countries or terrorist groups would choose different cities in the U.S. for "exemplary" attacks. Those coming most readily to mind are:

New York	Muslim extremists, Iranians
Washington	Iraqis, Serbs, Somalis,
Miami	Cubans, Palestinians
Los Angeles	Japanese

Keep current with the news that will provide you the latest list of potential targets. The most likely targets are:

- Population centers
- Transportation hubs
- Military bases
- Government offices
- Perceived centers of affluence
- Ethnic neighborhoods

Chemical warfare is the poor man's nuclear bomb. Almost any country or organization with a few trained technicians can turn out chemical warfare agents. A truckload of nerve gases could wipe out most metropolitan areas. A well-trained technician can even use a rail-marshaling yard as his mixing bowl to inflict damage on the surrounding area.

Chemical and biological weapons have a well-deserved bad reputation. They can't take orders once they are released. They travel with the wind and kill everything they come in contact with. You cannot surrender to them. They can't tell friend from foe. They have only one purpose—to kill. They reek of genocide by wiping an enemy off the face of the earth.

The existing binary nerve gases are technically not chemical weapons until the two components are mixed together. In binary agents, the mixing of two agents occurs as the bomb or artillery shell is being delivered to the target. Thus they are relatively safe to manufacture, store, and handle. Binary agents are less potent than primary nerve gases. For example, one canister of methylphosphoryldifluoride is mixed with a canister of isopropanol to get Sarin, or, as the armed forces call it, GB-2. Most binary weapons are made by the developed countries, but Iraq did have binary shells mixed prior to firing by the artillery crew.

The Gulf War exposed serious deficiencies in protective gear, gas masks, troop training, and supplies of antidotes. If Iraq had used its chemical arsenal, the result would have been a disaster for the coalition forces. We are facing a real threat to our armed forces, with an almost-defenseless civilian population.

Day One

In the armed forces, soldiers are trained to report any shells that do not explode upon impact. These might be "recycled" by terrorists for dispersing chemical or bacteriological agents. However, the landing of these shells would be on Day Two. What is Day One? Day One nears when you hear that nations are using chemical weapons in wars with their neighbors or that a terrorist organization is using them on its "enemies." You should have a gas mask for each member of your family. Ensure that they fit properly and have at least one spare filter for each. Filters must be changed if they get wet, so you may want more than one spare filter among your goods.

What to do:

- Have a bug-out kit with a gas mask and spare filters. If you expect the use of nerve gases, mustard agents, or arsenical agents, you should have a protective suit. A jury-rigged one can be made from a hip wader, raincoat, and plenty of duct tape.
- Obtain chemical-detection units. The simplest may be pads of chemical-detection paper.
- Have a gas mask for each member of your family, and remember the old rule—the best way to change a filter while under attack

is to don a spare mask and then change the filter on the mask you have removed. This does not apply to nerve gases or blister agents.

- Have on hand books dealing with chemical and biological survival and countermeasures.

Day Two

The use of chemical weapons will most likely be against larger population or military centers. It is unlikely that chemical weapons would be used against rural populations, except perhaps to poison the reservoirs of large cities. The first use of a chemical weapon will more than likely come without warning. After its use, there may be more than one group claiming credit, while the real culprit may remain quiet.

What to do:
- Start carrying your kit and gas mask with you.
- Be clean shaven. The seal between your face and the gas mask is impaired by facial hair.
- If you can, avoid target cities. If you must be in a target city, familiarize yourself with chemical weapons. If you know which type of chemical weapon could be used, that should tell you whether to avoid high or low ground. Chemical agents can be lighter or heavier than air and will settle accordingly.

Day Three

By this time, the nation will have undergone a chemical attack. This will give you an indication of the type of chemical agent or agents used, so read up on that agent and act accordingly. The government may or may not distribute gas masks to the civilian population. The sorry state of our civil defense network, coupled with the lack of chemical warfare protection devices for civilians, will leave most people fending for themselves.

What to do:
- Obtain antidotes for the chemical being used.

- Obtain strong bleaches, such as Super Tropical Bleach (STB), for decontamination of equipment and supplies.

Day Four

Maybe there will be follow-up strikes with chemical or biological weapons. If these continue, a pattern will emerge, revealing whether the strikes are against military or civilian targets. American response will be appropriate and overwhelming. However, this is little comfort to the ones hurt by the attacks.

What to do:

- Make sure that you have spare ABC-rated filters for your gas masks.
- Obtain additional decontamination supplies.

Day Five

If the attacks continue, the authorities will start to disperse the population by moving people from cities to the countryside. If the enemy has large stocks of chemical weapons, this measure will put the countryside in the target category as well.

What to do:

- If you can, move before the official evacuation. Moving early enables you to take most of your supplies with you. Forced evacuation usually allows you to move very little of your supplies.
- If you don't leave, stay out of sight. The troops remaining behind will shoot on sight to control looters.

Day Six

Regardless of whether the chemical attack is by a nation or a terrorist group, the U.S. will retaliate in a massive way. If terrorists were the perpetrators, we may see concentration camps for people with the same racial or ethnic background as them. This may even be akin to the concentration camps for Japanese-Americans during World War II.

If a nation is behind the chemical agent attacks, that nation as an independent entity will more than likely disappear from the face of the earth. America has demonstrated a number of times how bloody minded it can be about a sneak attack. It will be no different now.

What to do:
- Obtain more decontamination supplies if you live in a populated area.
- Make sure that your gas mask filter seals are intact. If damaged, replace them.

Day Seven

If chemical attacks continue, we are likely to see a permanent establishment of concentration camps for people originating in the country thought to be behind the attacks. This may lead to a terrorist movement among the heretofore uncommitted members of that ethnic group.

Even an inadvertent release of chemicals from a factory will be labeled as terrorism, leading to local hardship for the residents of the area.

What to do:
- If a concentration camp is located in your area, form a patrol force to ensure the safety of the neighborhood.
- Set up roadblocks to control the movement of people.

Day Eight

America will occupy the country behind the attack and will take action against any other country supplying equipment and raw materials for making chemical warfare agents. This will lead to an international enforcement of the ban on chemical weapons.

Since Germany, France, Switzerland, the Czech Republic, Russia, and the Netherlands are suppliers of chemical equipment and chemicals, we may see the United States taking punitive action against these countries. This will have a major impact on international trade and relations. We may see the European Community facing off against the U.S. Despite the E.U.'s stance on free trade, it is highly dependent on countries who are the potential chemical warfare aggressors against America.

What to do:

- Listen to shortwave broadcasts to obtain information on how American actions are viewed by other countries. This can alert you to potential retaliation against the U.S.
- In case of economic boycotts, supplying parts for equipment made in those countries will be very profitable.

Day Nine

Chemical weapons will be banned and not just in theory. An international force will inspect all nations capable of producing chemical agents. Many countries will be independent in name only. This may breathe new life into the UN.

Biological Warfare and Catastrophes

The use of biological agents as a military option is not likely. Notwithstanding, biological catastrophes have regularly decimated the earth's population in the past. Plague killed one-third of the European population in the fourteenth century, tuberculosis killed one-quarter of that same population in the nineteenth century, and tuberculosis is on the rise once again even in developed countries.

Recently, a variant of one of the three most deadly viruses ever discovered was found in the well-to-do Washington, D.C., suburb of Reston, Virginia. In its most deadly form, it killed nine out of ten people in African villages in Zaire twenty years ago. We have witnessed another outbreak recently in the same country. The variant strain showed up in monkeys imported for medical research. Researchers say it's just luck that it didn't jump to humans and mutate. The super-deadly strain, called Ebola Zaire, grimly boasts the highest mortality rate ever recorded for a human virus.

Mutant viruses will be one of the biggest problems we face in the next ten years. Viruses are now mutating much faster than they used to, and today's airline travelers unwittingly transport these deadly diseases all over the world—in a matter of weeks, days, even hours! Among the viruses, strange combinations switch genes back and forth. For example, China, where farmers raise pigs and ducks together, is a birthplace of new flu epidemics. The avian flu virus from the ducks infects the pigs. Once inside the pigs, the flu virus swaps genes with the mammalian viruses. This process creates new influenza strains that infect the farmers, thus requiring new vaccines to fight the outbreak.

A new plague strain of a common flu virus—airborne, acutely infectious, and more dangerous than AIDS—would take just seven days to cross the world and kill millions of people within a few days of becoming ill. Researchers say it's only a matter of time before another superdeadly strain, like the flu of 1918, mutates again. It is *natural and inevitable.*

Recently, we had a sobering warning when the Indian pneumonic plague nearly touched down on every continent in just two days. Globally, health officials went on full-scale alert. Passengers and cargo from India were refused in many countries. What if a terrorist purchased samples through legitimate research outlets and produced more in a clandestine laboratory? It can be done with a little preparation and by spending a little money. Just call the ordering outfit XYZ Biologicals Inc., and you could be in business.

The last flu epidemic (less than 100 years ago) was the world's worst plague. It's hard to imagine how quickly people died that summer and fall. You could wake up tired and achy on Monday and never live to see the week's end. The death toll was staggering. One out of seven people in Philadelphia and Baltimore died. This deadly

flu infected half the world's population and killed twenty-five million people, most within days of becoming ill.

There was no time to develop a vaccine. Death came too rapidly. Will it come again this winter—or next? Another grim area in this scenario is the reemergence of diseases once thought nearly extinct. Just look at the recent spread of tuberculosis, the flesh-eating bacteria, Legionnaires' disease, and hepatitis.

Add to these a number of nasties relatively unknown until transportation networks and social engineering allowed their wide-spread distribution. Some of these are the deer-tick-spread Lyme disease, the Tiger-mosquito-spread dengue fever and yellow fever, Hanta virus, and, of course, AIDS. As a microbiologist put it, "There is more genetic engineering taking place in your intestine than in all the biotechnology companies."

How and where do we get these diseases? Cholera from water, Legionnaires' disease from water vapor in air conditioning, influenza from airborne virus, *E. coli* from poorly cooked beef and raw milk. From insect bites, we can get malaria, Lyme disease, dengue fever, encephalitis, and yellow fever. From sexual activity, we can be exposed to hepatitis B, AIDS/HIV, and the traditional sexually transmitted diseases.

Consider the following example: The plague, *Yersinia pestis*, can be kept frozen for an indefinite time. What lies in wait for us in the antarctic ice sheet? Should global warming continue, more antarctic ice will slide in the oceans and eventually melt.

You are not even safe in a hospital. Hospital emergency departments are closing to combat bacterial infections. At one time, this was an American problem, but now it has spread. People so infected must be kept in isolation, and this is only the tip of the iceberg. What else is out there waiting for us?

On top of all of this, some 30,000 scientists of the Soviet Biopreparat program are hawking their skills on the world market. The Soviet Union had a twenty-five-year program to create a qualitatively new biotechnology. Imagine if Iraq had released its stockpile of *Clostridium botulinum, Bacillus anthracis,* and mixtures of fungal toxins. The shortages of gas masks, protective gear, and antidotes discovered during the Gulf War apply to biological agents, as well.

The problem of biological warfare is that if you don't know who has what, you do not know who threatens you. The lack of detectable armaments should result in the prudent assumption that you must consider all states to be potential targets. Even if you destroy the biological plants, they can be rebuilt quickly and cheaply. So, once again, the only deterrence will be the threat of massive nuclear attack on the state engaging in biological warfare. Even though biological weapons create a general equality among nations because the plants can be rebuilt quickly and one pound of starting material will very quickly create another pound of material, they are risky. First, there is the chance of the accidental release of biological agents. Second, the plants are prime targets (some Israelis call them "missile magnets") of an opponent.

Day One

The day is here now, today. Have no illusion about that. The "flesh-eating-bacteria" (necrotizing fasciitis) in less than twenty-four hours, if untreated, can be fatal. This bacterial disease, which kills connective tissue, presents an enigma. What turns a relatively benign Strep A infection into a deadly flesh-eating disease? We do not know. There are about 10,000 cases a year in the U.S. and about 2,000 of these patients die. With that kind of death ratio, it is little wonder that hospitals overreact.

With the U.S. well on the way to denying vaccinations to illegal immigrants, we are on the way to building a host pool for the reemergence of a number of diseases once thought to be under control. These can range from polio to smallpox.

What to do:

- Make sure that your immunizations are up to date.
- Refrain from having antibiotics prescribed for you for things like the common cold. First, they do not cure the cold. Second,

they give other bugs a chance to develop an antibiotic resistance in your body.

- Build up a good supply of antibiotics, and make sure to discard those that are beyond their expiration dates. You can lengthen the shelf life by storing them in dark, cool places. Crystalline penicillin has a long shelf life.
- Have a library pertaining to biological warfare and catastrophes. You should also subscribe to some general science magazines.
- Have gas masks, spare filters, and protective suits on hand.

Day Two

This is the day when an outbreak of disease is reported. Whether this is accidental or state sponsored is of little difference to you. You just want to survive it, not debate it. First you will hear some lurid stories, but very soon the authorities will start to manage the news. You will hear from official sources about a localized outbreak under control. Given the past performance of official news managers, if you believe them, you will also believe that banks provide a service and the oil companies compete.

One possibility could come from a recent news release in Australia. Rabbits have long been considered a pest on that continent, and now scientists have developed a virus that causes fatal internal bleeding in rabbits. What if the virus mutates and spreads to other animals or even human beings?

What to do:
- If there is a vaccine for the outbreak, get it and have your family immunized also.
- Keep away from public gatherings.
- Find out what the incubation period is for the disease. If you have visitors, isolate them from you for that duration, plus three days.
- Find out if the disease is waterborne. If it is, boil all the water you use for drinking, food preparation, and even for shaving.

- You should start to add to your supplies. If the disease is not brought under control, there will be shortages. If the disease is brought under control, you can always use your supplies later.
- Control fleas, rats, mice, squirrels, and vectors. Put flea collars on all pets. Spray all buildings to control fleas and ticks.
- Obtain duct tape, plastic sheeting, and other material to seal off your home. You should have an air filter for your home, otherwise you will have to wear your gas mask constantly.

Day Three

At this point, the hospitals are filling up, and now the medical personnel is exposed to whatever bug is doing the damage. The government will probably bring in medical personnel from other, unaffected areas, and police leaves will be canceled. This will result in nothing more than exposing additional people to the plague.

If the government kept the news of the outbreak a secret, it will become an item of rumor and may seem worse than it actually is.

What to do:
- If, in your estimation, the situation will worsen, evacuate if you can. If you can't, isolate yourself from the community. This is why you should have food reserves.
- For many of the plaguelike diseases, the dead must be cremated, and all items handled by the affected people must be sterilized.
- Remember, veterinary antibiotics work on people, too. Check the availability at local farms, farm supply stores, veterinarian offices, and other suppliers.

Day Four

The medical personnel are dropping like flies. The government will seal off the area to contain the epidemic. The National Guard will be called out to control looters and, more than likely, federal forces will be the ones used to seal off the area. All public services will likely grind to a halt. Lack of potable water, lack of sewer services, and uncollected garbage will add to the misery and help the spread of the disease.

Looters will operate practically unchecked, but they will pay for their travels by contracting the disease. That should reduce their numbers in short order.

What to do:
- Keep away from other people.
- Maintain a communications watch to determine the extent and spread of the disease.
- Seal off your house. Another wonderful use for duct tape.

Day Five

The cordoned-off area will be short of vaccines, medicines, and other medical necessities. The fear of contracting the disease will probably prevent resupply of the infected area. Urban areas will be short of food, further reducing the people's ability to resist infections. Hungry people will wander around trying to feed their families and will be exposed to the biological threat.

What to do:
- If you are in the affected area, lay low. Avoid becoming an unwitting victim by having to go out to buy food or anything else you may need.
- There are many herbs with antibiotic properties. See if any of these are growing in your area. Make use of them to expand your supply of medicines.

Day Six

There will be reports of the outbreak in other areas. This will lead to towns and villages cutting all contact with other parts of the country. Interstate and intrastate transportation will grind to a halt. This halt in transportation will cause widespread unemployment as plants run out of parts and supplies.

What to do:

- Listen to your radio and, in particular, to shortwave broadcasts of major international networks. Sometimes they have better information than local stations.
- This is your last chance to add to your supplies at a reasonable price. Stock up on items you use.

Day Seven

Paranoia will be high on the list of preventive measures. We will see the end of all public gatherings, the withdrawal of public services, which will lead to widespread shortages. Since no one knows who is infected or who is a carrier, mistrust in strangers will be high on the agenda of people living in any area.

The return to the walled city-state, like in the Middle Ages, will start. This isolationist behavior will create local shortages of all types of supplies.

What to do:

- You should have alternate systems in place for your water supply, waste-water disposal, and garbage removal.
- If you can, seal off your community from other people.

Day Eight

Around this time, if you are a traveler, you will not be welcome in most places. It will be very much like the Middle Ages during the plague outbreaks. That is, you will see walled-off villages, fear of strangers, and rampant superstition.

This isolation will result in widespread starvation, particularly in urban areas, and shortages of all kinds of supplies and equipment.

What to do:

- Establish a quarantine station for newcomers to your area. Even friends and family must be quarantined if they enter your area. The quarantine period will vary depending on the incubation period of the biological agents.

- You may want to establish some kind of barter networks with neighboring areas. There should be some kind of trade protocol to reduce or eliminate face-to-face contact.

Day Nine

A vaccine is developed. Now the problem will be the distribution network. In the meantime, people are still dying. Finally the spread of the disease will cease, and it is hopeful that we will learn from it. One thing is sure: Other bacteria and viruses will mutate to provide us with continuous challenges.

What to do:
- This is a good time to go into business to transport supplies and provide security for the transportation industry.

Water Wars

The California dream is ending partially due to lack of freshwater. The produce from that state is only possible because of irrigation. Texas recently saw a drought rivaling those of the "dirty thirties." Israel controls water wanted by Jordan, adding to the Middle East problems. Israel's other neighbors, Syria and Lebanon, are also in dire need of more water. It's possible that Mexico and the U.S. could clash over freshwater resources. To these, add India versus Bangladesh, Hungary versus Slovakia, and the proposed dams on the Euphrates and Tigris Rivers that would cause problems for the neighboring countries.

Much of the earth's surface is covered by water, but freshwater is only a small fraction of that. It is estimated that of the water on earth, 97 percent is seawater, two percent is bound in glaciers and the like, and another 0.5 percent is locked in aquifers too deep to reach. That leaves less than 0.5 percent accessible freshwater to drive our civilization.

The most likely water war scenario will be between the United States and Canada. The western states need water, in spite of the potential aquifer in Colorado's San Luis Valley having 120 million acre-feet of water. That aquifer has a potential value in excess of *$600 billion*. California needs close to 35 million acre-feet of water per year for irrigation. Yet there is need for more water. Where to get it? If you are in the western part of the U.S., Canada's British Columbia holds the answer. The water levels in Texas and other adjoining states are perilously low. Add to this the three-year drought in Mexico, and you have a scenario where America may have to take very arbitrary measures in regard to water diversion projects.

Irrigation is extremely water intensive. To grow a half ton of grain per person—enough grain to supply 50 percent of a person's diet for a year—can require as much as 400 cubic miles of freshwater per year. Water diversion for irrigation can have very devastating effects. In the case of the Aral Sea in Asia, this resulted in lower sea levels, dying basin wildlife, salt buildup, and increased wind erosion.

Other problems concerning potable water supplies are caused by population explosion, aging water systems, and growing resistance of microorganisms to water treatment chemicals. Potable water will become a resource much like oil. Until now, we have been worried about pollution by chemicals and metals, but in fact, the problems increasingly are infectious diseases. Up to 80 percent of these diseases are caused by waterborne microbes.

Microbe detection methods are improving greatly, and active surveillance of disease outbreaks is better. However, new testing methods have revealed that many more microbes are waterborne than had been known previously.

Day One

Once you hear that the U.S. supports Quebec separating from Canada, Day One has already passed you by. Day One starts with the curtailment of water usage in California, Colorado, or other states. Another sign is new types of agricultural products coming from the affected areas. Shallow root crops like lettuce, potatoes, and tomatoes may give way to crops with deeper roots, such as corn, alfalfa, and grapes.

Unless new water sources or water conservation measures are found, America will face some unpalatable choices. Some of these are:

- How much water should a state like California get as opposed to a state like Arizona?
- How should water be allocated within a state? For example, is it better to allocate water to labor-intensive crops such as lettuce and thereby stave off hardship and unemployment, or to agribusiness and thereby make the most efficient use of available water?
- When does our good neighbor policy end? Northern Mexico's agriculture is critically dependent on American water. Just how can we turn down our tap?

The answer to these questions will determine what America will be like.

What to do:

- Keep up to date on what is happening around you.
- Take a water inventory of your area. Get to know where your water supplies come from and where there are alternate supplies, if any.
- Review your water consumption, and reduce or reuse water wherever possible.

Day Two

You will see the government finance schemes like bringing icebergs from Antarctica to Los Angeles. The water levels in the Great Lakes will continue to drop, according to some scientists, as much as 25 percent.

Crash courses in water reuse will be given to employees of those organizations that are large consumers of water. New water-saving appliances will be mandated by some states.

What to do:

- If you expect water shortages, invest in local water supplies.
- If you live in the country, have another well drilled in a secluded area to use as a backup in case your primary well is controlled by others or is regulated by the authorities.

Day Three

Water conservation measures will be introduced. There will be very heavy fines and jail sentences handed out to polluters of the groundwater resources. Price increases for foodstuff grown in the most affected areas will start a trend toward much higher food prices nationally.

Some corporations will move water-intensive activities to other countries, adding to local unemployment. This will start a stampede of industrial activity moving from the affected area.

What to do:

- You can make a living if you set up a testing laboratory for water and water contaminants.
- Try to obtain hydrogeological maps of your area.

Day Four

If the above measures do not work, additional water conservation measures will be introduced. The authorities will search for additional freshwater supplies. Food prices will keep escalating, particularly fresh vegetables and rice. These price hikes will cause unrest, especially in poorer inner-city areas, and local violence will increase.

What to do:

- Invest in hydroponic farming operations. You may even want to start a small operation yourself.
- Find a reliable dowser. Consider going into the well-drilling business.

Day Five

Areas where agriculture is dependent on irrigation will be greatly affected at this point. Food prices will rise further, and there will be shortages of certain commodities. In response to this, the government may decree that use of grains for alcohol production be prohibited, and production of tobacco may be just a memory. It may come to the point that cattle will have to feed on pasture only. Get ready for gamy meat.

More than likely, a population shift will occur as a result of water shortages. The movement to so-called water-rich areas will accelerate as the water shortages become chronic.

What to do:
- A cooperative for food procurement and production may be the best way to go at this time.
- If your area is subject to population movements, evaluate what effect the reduced population will have on your future. You may be much better off staying put.

Day Six

Water-recycling measures will be introduced. Most of these will be in agriculture and industry. Consequently, people taking shortcuts will contaminate food and goods. This will lead to further government controls and inspections. The inspection teams will be given powers only seen under an occupation or martial law scenario.

What to do:
- The age-old method of rainwater cisterns should be evaluated. This depends on the precipitation in your area.
- Get additional information on your area's water resources. Keep track of changes in your area aquifers.

Day Seven

"Temporary" water curtailment for residential purposes will be introduced. If you have a good-looking lawn, questions will be asked. In the meantime, the search for additional water sources will continue. Inefficient reuse of water will result in contamination of food-processing plants, adding to the problems.

What to do:
- Add additional water-purification equipment to your supplies.
- You can store water by adding a small amount of bleach to it.

Day Eight

Parts of Canada and Mexico will be occupied and large-scale water diversion schemes will begin. Depending on the occupied people's cooperation, or lack of it, the work pace may be slow or fast. In the meantime, the water shortages will continue. Many people will be employed on these water-diversification projects. This full employment will provide the government with a partial justification for the occupation.

If there is violent resistance on the part of the occupied population, even more people will be employed by the armed forces and security agencies.

What to do:
- This is a good time to draw up long-term plans for maintaining your self-sufficiency of freshwater supplies. Your children and grandchildren will thank you.

Day Nine

Now that new water is available, the question is whether we return to our previous consumption patterns or we have learned to reduce, reuse, and recycle.

Weather Wars

The weather we call normal is, in fact, highly abnormal. For the past seventy years, we have enjoyed some of the best weather mankind has ever known. There is a growing consensus among climatologists that the world is undergoing a cooling trend in spite of the "greenhouse effect."

There is also human intervention. Imagine that a hostile power seeded clouds with silver iodide, and as a result, all rain falls on barren ground. Your prime crop-growing region is an arid wasteland. (Of course, the clouds have to be present before they can be seeded.) Or even worse, imagine they managed to divert the monsoon rains from India and Southeast Asia, putting more than a billion people at risk of imminent starvation. Far-fetched? Perhaps, but possible.

In addition to state-sanctioned weather-war incidents, we also have local events. These can range from a dam being erected in a unique microclimatic area to a program for straightening a river in order to control floods. Moreover, we now realize that for the past seventy years, we have had abnormal weather in the sense that our normal weather is cooler and prone to more storms than we have had lately. Now that things are returning to normal, it will be harder on us. When this happens, the politics of food will become the central issue for every government.

Precipitation patterns are so important that some geographers equate the significance of the precipitation map with that of the world-population density map. To produce two tons of dry matter (grain) per acre, it will take anywhere between 800 and 1,000 tons of water.

The possibility of weather as a weapon has been studied by the CIA as far back as the '60s. One of its reports stated, "In the poor and powerless areas, the population would have to drop to levels that could be supported. Food subsidies and external aid, however

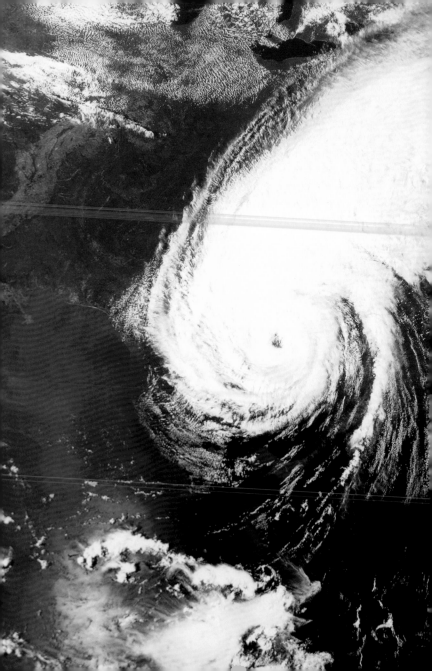

generous the donors might be, would be inadequate. Unless or until the climate improved and agricultural techniques changed sufficiently, population levels projected for the LDCs (less-developed countries) could not be reached. The population 'problem' would have solved itself in the most unpleasant fashion."

Now the U.S. has the HAARP (High-frequency Active Auroral Research Program). In addition to jamming global communications systems and disrupting animal and human behavior, this is said to be able to change weather patterns over large areas. It appears to have the ability to impact the earth's upper atmosphere. Talk about a scary weapon! This project is probably the culmination of a recommendation made by the World Health Organization and the World Meteorological Organization in 1976. The two organizations recommended that all nations study the possible action of various parameters of natural electricity on man.

Ultimately, it is unimportant whether a weather change comes as a result of human intervention or from natural causes. The change will result in wars. Can you imagine China feeding almost five people per acre? They do *that* now. A weather change would have the Chinese expanding in all directions to feed more than one billion people.

Day One

The Russians undertook many major studies pertaining to the use of weather as a weapon of war. These ranged from cloud seeding with chemicals to using high-altitude explosions to change wind patterns. Whether they, the HAARP, or Mother Nature decide to play havoc with the mild weather we are experiencing is immaterial. What is sure to come is a change in the weather as we know it. As said many times before, you can't predict an ice age until you are in it. Then the question is whether it is a little or large ice age. A little ice age affected the world from the 1600s to about 1850.

Professor Reid A. Bryson, while at the University of Wisconsin, put forward the most likely impact of a little ice age. These would be:

- More rain would fall in the northern United States.

- The central Gulf Coast, the Southwest, and the Rockies would become drier.
- The winter-wheat region of the High Plains would be much wetter, although yields would not be affected.
- India would have a major drought every four years, and more than 50 million metric tons would be required from the world's grain reserves to prevent the deaths of 200 million Indians.
- China would have starvation conditions every five years and need 100 million metric tons of grain for its people. Cooling in the north and a reduction in the monsoons in the south would deal a double whammy to China.
- Canada would lose 50 percent of its food production and reduce its exports by 75 percent.
- Northern Europe would live in a permanent twilight of winter. Even the summers would be cool and dry, and harvests would be poor. Winters would be snowy and bitterly cold.

How does this process work? Up to about 1940, the world warmed; then from about the midforties on, a reversal took place as polar air expanded south and the Northern Hemisphere cooled. The three main trends involved are volcanic dust, man-made dust, and carbon dioxide. Now volcanoes are erupting all over the place, and man-made carbon dioxide emissions have increased, starting the cycle leading to an ice age. As little as a 2.8°F drop in global temperatures would start the glaciers marching.

What about a great ice age lasting some 90,000 years? During the last great ice age, all of Canada and the northern U.S. were covered by glaciers. In Europe, the ice sheets went as far as Britain, Germany, Switzerland, and Holland. Australia's outback was cool and wet. Much of Chile and Argentina was icebound.

What to do:
- As always, have your emergency kit ready to go.

- Have a food reserve. If you can afford it, the food reserve should be large enough to feed all members of your family for a year or more.
- Make arrangements to reduce your need to visit government offices and installations.
- Obtain up-to-date information on these threats; reading scientific journals is the place to start. If you are not into heavy science, read back issues of *Popular Science* and like publications.

Day Two

We may start to realize that it is not Mother Nature, but man-made interference that is causing our weather extremes. This will lead to a search for the culprits, who we may or may not identify at this early point. If the weather changes are natural, the search will focus on what triggered them and how to stop them.

What to do:
- Depending on the nature of the threat, add appropriate supplies to your reserves.
- This is an excellent time to move if you know what is coming. Southern real estate prices will be reasonable at this point. On the negative side, if the weather changes are temporary, you may lose money on a quick sale.

Day Three

The changes in the weather will most likely impact food production. At this point, you may know whether the weather changes are temporary or permanent. Meanwhile, famine will threaten millions along the monsoon belt. A cycle of droughts and floods, hurricanes and tornados, killing frosts, and avalanches will be common. They will occur far more often and far more randomly than ever before.

Militarily strong but hungry nations will mount invasions. Massive migrations backed by force will start.

What to do:
- Lay in supplies of seeds, but make sure they are not hybrid seeds. Store them in sealed, carbon dioxide- or nitrogen-filled cans.
- Keep a journal of local weather events, and compare these with the historical record. Remember, one unusual year does not mean a weather change. It may be only a blip.

Day Four

We will retaliate. Weather is no respecter of national boundaries, so we may be victimized by our own retaliation. If the weather changes are due to natural causes, we may see crash measures taken to reduce the dust and carbon dioxide levels in the atmosphere.

The crop strains of the Green Revolution will be badly hit. They are dependent on the abnormally warm weather of the 1950s and 1960s and offer less resistance than older, indigenous strains. They cannot survive outside the narrow spectrum of our abnormally warm temperatures and rainfall. This, on top of reduction in farmland due to urbanization and land erosion, will lead to increasing food shortages. China and India already practice double and triple cropping.

To add more arable land, the government will undertake massive projects of river diversion, swamp drainage, dams, and other schemes. All of these will be useless if the weather takes another random turn. The food-exporting countries—the U.S., Canada, Australia, Argentina, and France—will be forced to focus on feeding their own people first, leaving precious little for food aid to other countries.

What to do:
- It is time to get into food production. Whether you do it as a farmer or as a member of a farm cooperative is not as important as the realization that you must be self-sufficient in food.

- Obtain older books dealing with agricultural practices. Those from the early twentieth century will be very useful in the coming days.

Day Five

The pressure on food production will increase. Reduction in farm yields will result in shortages and higher prices. Some spectacular failures in hybrid seeds will occur. The crop failures will add to a worldwide recession and will cause many border conflicts to erupt.

What to do:

- This is the time to decide whether the changes are permanent. If you feel that they are, you must evaluate whether the area you live in is capable of supporting you and your family.

Day Six

By this time, weather extremes will be the norm. The "once-in-a-hundred-year" floods will happen every other year or so and will place enormous stress on agriculture and transportation services. Around this time, only the U.S., Canada, and Australia will be food-exporting nations. The politics of food will be more and more important.

In the previous little ice age the population was small, so small tribes simply moved south. Now we have billions who would be on the move. Most of these would be urbanized populations who know little about agriculture and would make lousy nomads. They have neither the physique nor the training to be of much use when they pour out of the cities.

What to do:

- Weather extremes will force you to review your living patterns. Be ready to change them. For example, if you have a solarium and now you experience hailstorms every year or so, it may be prudent to cover the glass with plywood during certain seasons.

- The idea of fish farming as a way of obtaining protein should be considered. Fish convert meal to protein more efficiently than land animals do.

Day Seven

The authorities may decide to concentrate people in "safe areas." In addition, food distribution will only be available to those residing in these areas. These will be nothing more than glorified concentration camps. Avoid them. The areas where the camps are located will be heavily patrolled to nab escapees and stragglers.

What to do:

- Find out which areas are affected by the weather changes and the extent of these changes. Also, locate areas with suitable microclimates, and see if you can move there.
- Avoid joining any unorganized groups. They can't help you. Each person is on his/her own.
- If you live in an area where camps are located, think about moving. You may lose some of your assets, but save your life and liberty in the process.

Day Eight

We learn to live with the weather extremes. This may result in using seeds from times before the hybrid-seed development efforts. The price of food will rise to the point that most families will have to spend more than 60 percent of their earnings on food. More and more marginal land will be put to agricultural use.

What to do:

- Maintain a radio watch to alert you as additional weather changes occur, and keep track of new weather patterns.

- Measure the pH of your soil to find out if it is acidic or alkaline, and combine this information with the weather changes to figure out which plants will grow best.

Day Nine

Now is the time that the extent of the weather changes will become evident. Even more importantly, you will see if these changes are permanent. If we are in an ice age, massive southward migration will be taking place. This migration will displace lesser-armed populations, and the toll in terms of life will increase.

More and more of our children have asthma, allergies, and other disabilities traceable to pollution and the stress of modern life. In some ways, we are beginning to see that trace quantities of myriad chemicals in our air, land, and water have cumulative effects. The problem lies in the fact that once something is invented, it is impossible to uninvent it. Similarly, we have come to depend on the chemicals available for all kinds of applications. These, coupled with our learning too late about the long-term side effects, create a major hazard to our continued survival as a species. Taken to its extreme, we may end up living in hermetically sealed bubbles just to exist.

The human species is capable of altering the ecological balance of the planet and has been doing so big time for the past century. With increasing population, we are clearing land by burning rain forests, reducing floods by straightening rivers and streams, and cultivating the wrong crops in different lands. The soil nutrients are depleted. Biodiversity is at risk while we are continuing our search for the perfect hamburger.

Added to this are old-fashioned events, ranging from fires to famine. We have had plenty of experience dealing with ecological disasters, but are, nonetheless, still ill-prepared to do so. The population explosion adds to ecological problems. If the earth's population would drop to 10 percent of the existing number, humans would have very little impact on the ecology of the planet.

6

Ecological Survival

Another problem in common with other scenarios is our inability to forecast the impact our actions will have on the ecosystem. So we end up adding more chemicals to cure a problem caused by chemicals in the first place. The plant kingdom is a miniature chemical factory producing far more chemicals than animals, and as a result, plants have been relatively immune to all but the most obnoxious of our pollutants. That is not so now. We are not quite sure what changes we have already caused. We may have a ticking ecological time bomb out there affecting plants as well as humans.

Day One

Day One arrived some time ago, although no one is quite sure if it was in the late 1980s or the early 1990s. Codfish have been virtually eliminated in parts of the North Atlantic. There was Chernobyl, chemical discharges from industrial plants, gold mining in developing nations, and smog alerts in major cities during summer. Mankind has had a major impact on the world's ecosystem. Whether the earth's ecosystem can make corrections in time is the question facing us.

The most likely of these ecological scenarios is the food wars. Famine is with us at all times in one part of the globe or another. It can come from weather, transportation, politics, or other causes. The results are the same—people starve.

What to do:

- As always, have your bug-out kit ready.
- Update your skills in first aid, CPR, and other lifesaving measures.
- Have medical supplies on hand.
- Read about measures taken in previous ecological disasters. This should give you an idea of what will work and what to expect.
- You may want to invest in stocks of environmental companies. You could really clean up.

Day Two

This is the phase when you decide what kind of scenario you are facing and read the appropriate section in this book. There is a certain overlap between ecological and climatic scenarios.

Chemical Spills and Evacuations

Chemical spills are more and more common nowadays. Even major population centers have been evacuated in the last few years. One of the largest was Mississauga, a city adjacent to Toronto, Ontario. More than half a million people had to leave their homes. The major transportation networks connect major cities. That is what they are supposed to do. This places all those who live close to transportation routes, plants, and industrial centers at risk of being uprooted at a moment's notice.

Everyday life in modern times is full of potential hazards. At one time, Dupont's advertising stated, "Better things through chemistry." Unfortunately, there is also a downside to this state of affairs. Modern industrial processes require large amounts of chemicals, many of them hazardous to your continued well-being. These chemicals must be transported cheaply and efficiently from the point of manufacture to the point of use. Therein lies the danger.

Our transportation networks were mostly laid out in the nineteenth century or even earlier. Given this, the interchange hubs are often located in the middle of our largest cities. This can lead to a train derailment necessitating the evacuation of a million people. For various reasons, you may want to wait it out in your home rather than evacuate. Thus, you must be familiar with protection measures.

First and most important is a source of current and local information. This must include scanner radios. Up-to-date information

is extremely important in order to make a stay-or-go decision. The information should include data on the weather, including wind direction and speed.

Let us review the commonly encountered dangerous chemicals and their properties.

Ammonia (NH_3)—Aside from its pungent odor, the most striking characteristic of ammonia is its extreme solubility. It is even more soluble in water than hydrochloric acid. It is about one-half as heavy as air. It is easily liquefied at ordinary temperatures at nine atmospheres (132 pounds per square inch). Liquid ammonia is colorless. It boils at −33.5°C (−29°F) and freezes to a colorless crystalline solid at −78°C (−110°F).

Liquid ammonia is extensively used in refrigeration systems and in making artificial ice.

Chlorine (Cl)—Chlorine is a greenish-yellow gas with a very irritating odor. It is extremely poisonous and exposure, even for short periods to a low concentration of the gas, will produce serious lung problems. It is about two-and-a-half times as heavy as air. It can be readily liquefied at ordinary temperatures by applying a pressure of between six and seven atmospheres. In this form, it is an article of commerce. This is the usual form for transportation. It boils at −34.5°C (−30°F) and freezes at −101°C (−149°F).

Chlorine is extensively used as a bleaching agent in the pulp and paper industry.

Formaldehyde (HCHO)—This is a gas with a penetrating odor that causes the eyes to smart. It causes irritation of skin, eyes, nose, and throat. It is used in a water solution as a disinfectant and an antiseptic. Its other uses are in making resins, chemicals, and dyes. It is used as a preservative, reducing agent, corrosion inhibitor, and in the recovery of precious metals.

Hydrochloric Acid (HCl)—The solution of hydrogen chloride in water is known as hydrochloric acid. Concentrated hydrochloric acid is a fuming, colorless liquid. It is about 20 percent heavier than water and causes burns on contact with skin.

Hydrogen Peroxide (H_2O_2)—Concentrated hydrogen peroxide is a strong oxidant. Concentrations of more than 65 percent present a fire and explosion hazard. It is used for bleaching, dyeing, as a rocket-fuel oxidizer, as an antiseptic, and many other chemical formulations.

Nitric Acid (HNO_3)—Pure nitric acid is a colorless liquid. It fumes strongly when exposed to moist air. It can be mixed with water in all proportions. It is highly corrosive—just one drop on the skin causes a bad burn. Even the diluted acid will turn clothing and skin yellow. To decompose nitric acid, dilute with water and expose it to sunlight. This is a slow process. When the acid is heated, the decomposition is very rapid.

Sulfur (S)—Sulfur is a pale-yellow, brittle solid and is about twice as heavy as water. It is not soluble in water and has no marked taste or odor. It boils at 445°C (830°F) and upon burning forms sulfur dioxide. This gas has a choking effect and is more than twice as heavy as air. If water is used to put out a sulfur fire, the sulfur dioxide forms sulfurous acid (H_2SO_3), a milder acid than sulfuric acid.

Sulfuric Acid (H_2SO_4)—Concentrated sulfuric acid is a colorless, oily liquid about twice as heavy as water. It boils at a high temperature (338°C). It is an exceedingly corrosive liquid and when spilled on the skin, produces bad burns. It mixes with water in all proportions, producing a large amount of heat. Since the concentrated acid has a great attraction for water, it is used for drying many gases. The hot concentrated acid is an oxidizing agent. Sulfuric acid is used in huge quantities. It is the heavy chemical that enters, in one way or another, into the manufacture of nearly everything we use.

Vinyl Chloride—The most important of the vinyl monomers, this extremely flammable gas and liquid (under pressure) is used in most plastic syntheses.

Day One

Assume that a chemical spill and consequent evacuation can happen to you at any time. Protective measures must be taken now.

If you live close to a transportation hub, you should have a bug-out kit with you at all times.

What to do:
- Have a small emergency kit prepared.
- Have a gas mask and spare canisters for each member of the family. The best way to change canisters is by having a spare gas mask. Switch masks and then change the filter on the gas mask that has been removed.
- Know how to protect against chemicals by reading the appropriate reference books.

Day Two

If you believe that a chemical emergency has occurred, but the area has not been evacuated, listen to your scanner radio to find out what is happening. It very well could be that the event is under control. On the other hand, if you hear that a propane car is on fire, you must move fast.

What to do:
- Go indoors, and stay there.
- Close all windows and doors.
- Turn off the furnace or air conditioning and any fans.
- Close drapes and fireplace dampers.
- Place moist towels at the base of doors to act as an air seal.
- Fill pails, containers, and the bathtub with tap water.
- Listen to local broadcasts. If you have a scanner radio, use it to monitor official emergency communications.
- Be prepared to move to upper or lower levels of the building if instructed to do so over the radio or by emergency officials.
- If breathing becomes difficult, use a gas mask or breathe through a moist towel.
- If you need immediate help, place a white towel or cloth in a window facing the street, or flash a light on and off in the window. Call the fire department or 911.

- Do not leave the building until emergency officials say the area is clear.

If you are asked by emergency officials to evacuate, you, your family, and friends should have a predetermined meeting place. The preselected meeting place should be well away from potential chemical spill sites.

What to do:
- Take your emergency kit with you.
- Drive to your predetermined evacuation meeting place or to the evacuation center announced over the radio by emergency officials.
- If you do not have a car, follow instructions given over the radio.
- If you go to an evacuation center, telephone those friends with whom you arranged to have a predetermined meeting place.
- If you evacuate to your predetermined meeting place, have a communication system to keep in touch with other members of your party.

Day Three

This is when you return home or when the officials declare that the emergency is over. Your home may be looted or ransacked.

What to do:
- Find out what damage, if any, has occurred to your home.
- If curtains and clothing smell of chemicals, remove them and air them out outside.

Day Four

This is the time to call insurance adjusters and lawyers. By now you will realize that none of us is safe anymore. Think about the probability of the same thing happening in the future.

What to do:

- Take photographs of all damage to your residence. Prepare detailed lists of damaged items. Be sure to include serial numbers, if any, date of purchase, name of the seller, and original purchase price.

Fires

When a fire engulfs your home, you suffer more than just monetary loss. You also lose your sense of security. Most people who go through a home fire have their personalities deeply marked by the event. The best way to avoid fires is to make sure they cannot start. Buying chintzy curtains of an unknown material at a garage sale is almost as bad as keeping a can of gasoline in your basement.

You and your family should have a fire evacuation plan in place. This plan should identify a meeting place for all of you. Everyone should agree beforehand who should remove what from the house.

Then there are forest fires, refinery fires, high-rise fires, and a host of other possible fiery events depending on where you live.

Day One

Day One is now. You never know when a fire may break out. Fire can start in many places. One way to reduce the chances of a fire is to keep combustible material away from potentially hazardous places.

What to do:

- Obtain fireproof file cabinets or safes. Often you can find them in used office-equipment stores.
- Have a small emergency kit ready.
- Purchase several carbon dioxide or dry chemical fire extinguishers. Make sure that they are periodically inspected.

- Install smoke detectors and alarms in your house and business. Be sure to change batteries at least annually.

- Photograph the contents of your home, and have these photos in a safety deposit box or in a fireproof filing cabinet where you work. List all items, where you bought them, and how much you paid for them. Do not keep these lists at home as they may just become additional combustibles. Keep your lists up to date.

- All family members should sleep with the bedroom doors closed. A closed door can delay the spread of fire with its accompanying smoke and gases. Windows should be partially open to increase the oxygen supply in the room.

- Clothing fires are a sure way to get hurt. Buy only clothing meeting flammability tests.

- Have secondary exits designated for each room.

- Purchase an escape ladder for second-story bedrooms.

- If you live or work in a high-rise building, you may want to invest in a disposable escape hood.

- Have a fire drill at home.

- Have a pair of shoes and a flashlight under your bed. Check batteries every three months. These will come in handy during a fire.

- Arrange with a friend or neighbor to take in your children if you are not at home.

Day Two

This is the day when your house or business burns down. Teach all family members how to test for fire outside their rooms. Placing a hand on the doorknob or a door panel for six to ten seconds to see if it is warm will tell them whether to use a secondary exit.

What to do:

- Evacuate the burning building. Do not try to fight a large or rapidly spreading fire.

- If you can, remove flammables, important papers, firearms, and ammunition.
- Make sure that family members stay out of the burning structure.

Then there are forest fires. You may be alerted by the authorities that a forest fire is on its way, or your only warning may be the smell of smoke. You may see animals fleeing or behaving in unusual ways.

What to do:

- Do not flee at once unless the fire is at your camp. Plan your escape route. The smoke will show you the wind direction. The fire will spread primarily in the direction of the wind.
- Keep your clothing on. It will shield you from the full force of the radiated heat.
- Head for any natural firebreak. This can be a stream, river, open field, or road.
- Fire goes faster uphill, so do not head for high ground.
- If you have time, you can fight fire with fire by creating a firebreak.

Day Three

This is the day after the fire. If your building is burned out, there is not much to do but call in a bulldozer. But in case of lesser damage, you may be able to salvage many items.

What to do:

- Have someone keep watch on the burned-down building to prevent looting.
- Have your insurance adjuster visit the building.
- Start to prepare lists of what was destroyed in the fire. Do not forget to have backup documentation to prove how much you paid for those items.

Day Four

Now you are awaiting your insurance money, if you're getting any. Remember, insurance companies will try to pay you less than full value. This is called "low-balling." Find out what recourse you have against this practice in your state. Do not settle for the initial offer— challenge the insurance company's estimate item by item.

What to do:

- Make a complete list of what is destroyed or damaged, and get estimates for the repairs or rebuilding.

Poisoned Air, Lands, and Waters

We have heard about AIDS and other viral diseases breaking down the immune system. To these you have to add ultraviolet light and chemicals. Toxic substances that break down your immune system can kill you. We have dumped enough toxic substances into the environment that many people are writing end-of-the-world scenarios around them.

Sometimes a little pollution is beneficial, as was the case with sulfur dioxide in England. After emissions were greatly reduced there, farmers found stunted growth in cereal crops. Once this was identified, the farmers had to add about twenty kilograms of sulfur per acre per year to the soil to maintain soil productivity.

At the other extreme, mercury in fish, chlorofluorocarbons in the upper atmosphere, and heavy metals in the soil pose major problems to our continued well-being. It would be easy to advise moving to unpolluted areas, but the need to earn a living may preclude such a move. Even moving may not solve the problem, since blighted lands have a habit of shifting with time. You may have a country home fifty miles from the nearest industrial activity, and yet a check of your well water may disclose the presence of large quantities of heavy metals and dry-cleaning solvents. What now? At the very least, you will end up having to install a filtration or chelating system. Then you have the additional problem of disposing of the filtrate or the chelate in such a manner as to preclude it from re-entering your food or water supply.

On top of these perils, there has been a steady reduction in sperm levels in European males. I'm sure if worldwide studies were done, this would be a worldwide concern. Pesticides and insecticides act like hormone modifiers in higher order animals. We have documented changes in many birds, fish, and even mammals.

All these threats come at a time when we are concerned over the planet's food-production ability to feed the ever-increasing population. A ban on the consumption of fish from many waterways, at the same time requiring additional food sources to feed people living along those waterways, is a surefire formula for disaster.

Day One

When the U.S.S.R. broke up, we got a glimpse of ravaged lands, lakes dying, radioactive wastelands, and huge plants with no air scrubbers. And these were just a few of the environmental nightmares that came to light. Meanwhile, in America, now that many years have passed, all the used dry-cleaning solvent containers are leaking their poison into the groundwater. Closed service stations pose another source of pollution for the groundwater.

Now we have the Gulf War syndrome, this coupled with the widespread use of depleted uranium (DU) projectiles and tank armor. DU was first used on a large scale in Desert Storm, some 350 tons were used and are still in that area as either projectiles or fragments. Some of these particles are as small as five microns in size. Once inside the human body, uranium particles tend to stay, causing illnesses that often take decades to manifest, such as lung cancer and kidney disease. A single uranium particle trapped in the body

can expose surrounding lung tissue to approximately 1,360 rem per year. That is some 8,000 times the radiation dosage for whole body exposure allowed by federal regulations for the general public.

What to do:

- Keep an active interest in what is happening in your nation and your region.
- Have an emergency kit on hand.
- Research the history of any piece of property you own, rent, or plan to use. You may find that it was once a toxic waste dump.

Day Two

This is the day when some obscure scientist will bring to the public's attention that a longtime practice or a large employer poisoned part of the biosphere and your children may be debilitated as a result.

For all we know, it may start from a simple misprint in a user's manual. Instead of saying "change batteries in the wireless," it might read "change wires in the batteryless," and have wide-ranging repercussions.

What to do:

- Get as much information as you can. Moving too hastily can be detrimental to your health and pocketbook, but doing something too late is equally detrimental. You will have to judge the timing carefully.
- If some food production is affected by the event, stock up on that food. You may even want to use some of your surplus for barter.

Day Three

The government is starting commissions and studies to determine if the obscure scientist has a case. In the meantime, he is being hung out to public ridicule by his peers. No action is taken by the

government. The tabloids will have a field day, and if the poor whistle-blowing scientist is interviewed by them, his credibility will suffer even more.

This is the denial stage in the event's unfolding. Although there is much publicity, the government is inactive and tries to create the image of business as usual.

What to do:

- Keep up to date on the changes, and follow the developments. If you have friends in the scientific community, get them to follow up on the events and translate the findings into plain English.
- Discuss with your friends what you can do if the trend continues.

Day Four

The scientist is vindicated, but now it is too late. Permanent damage has occurred. There will be a crash program to rectify the situation, but more than likely, it will not work as planned. The research institutes favored by the government will take a plodding, step-by-step approach that will produce no immediate results. Crackpot solutions will be advocated and evaluating them will add to the time wasted.

What to do:

- This may provide you with a business opportunity. By having information on the nature of the threat, you may profit by supplying services or equipment for the crash program.
- Keep track of other ideas on what caused the problem, even if they sound preposterous at first. Some may be proven right.

Day Five

The government will overreact (what else is new?) and will ban many products and technologies from the market. Different governments

will ban different items, creating an opportunity for smugglers and black marketeers.

The black markets will give rise to another federal effort, which may be called the "War on Smuggling." This will lead to excesses on the part of the federal agencies involved.

What to do:
- Take a sober look at the newly introduced government controls. If they are cosmetic, ignore them in a circumspect way.
- You may be able to deal in contraband simply by repackaging it. This can take the form of labeling carbon tetrachloride as fire-extinguishing liquid or a metal degreaser.

Day Six

The search for the guilty party will focus on knowledge industries. These are universities, private development laboratories, and research centers. The manufacturers will claim that inadequate testing led to all the problems, and politicians in need of campaign funds will listen to them.

What to do:
- More business opportunities in the testing business.

Day Seven

The scientists will be found guilty of harming the planet. A new society will be in the making through central control of all scientific research. This may lead to a cessation of progress in the sciences. There will be many unemployed scientists. We will have, in one step, become Luddites. These bands of English rioters, organized to destroy machinery, first appeared in Nottingham and the neighboring districts toward the end of 1811. This was in reaction to the introduction of textile machinery, which put many craftsmen out of work.

What to do:
- Obtain formularies to prepare chemical compounds. Over-reaction by the government will result in shortages of many everyday chemicals.

Day Eight

People will turn on the scientists, many of whom will pay with their lives. The authorities will follow this hatred of science with an official rejection of science. This is very unfortunate and will result in the loss of much of the hard-won science we have today. Even worse, this will stop research for solutions to the many problems we have.

What to do:
- If you can employ scientists, you may have a whole new business on your hands. Those who turn their back on science will lack essential supplies.

Day Nine

We will see the emergence of a stagnating society. More and more people will be needed simply to produce food, resulting in the depopulation of cities. We may even see the return of animal power for farming instead of being blessed by the tractor.

Food Wars

The United Nations' FAO (Food and Agriculture Organization) released a report in late 1996 calling for a 75-percent increase in food production in the next thirty years. The "green revolution" in the 1960s doubled food production. The use of hybrid plants, fertilizers, and pesticides fueled this production increase. The downside has been the nitrification of rivers, reduced biodiversity due to inappropriate use of pesticides, and a dependence on petroleum products. The high cost of fertilizers has denied much of this food-production improvement to the African continent.

One of the side effects of cultural practices and population growth has been that some formerly food-exporting nations have become food importers. Take the Republic of Guinea on the west coast of Africa as a prime example. When the republic declared its independence from France, the saying in the country was "no meal is complete without rice." Denuding the hillsides for firewood leads to soil erosion, trying to plant crops in the leached soil leads to crop failures, and the resulting migration to urban centers leads to poverty. Today Guinea is a food importer and another economic basket case, as is much of Africa.

At the start of the twenty-first century, in much of the world, the only thing between starvation and survival is the timely arrival of cheap (sometimes free) food aid from America. What if America had to curtail food exports? Within a year of such curtailment, we would see most nations at war! Some of these countries, formerly recipients of free food from America, would declare war on us or foment terrorist campaigns against us. That's gratitude for you.

Another possible cause of food shortages has to do with the green revolution. Our scientists came up with many hybrid seeds that give greater yields in poorer soils, but the downside is that most of these hybrids are more susceptible to predators, fungal growth, and the like. The priority during the development of new seeds was on yield, not resistance to pests or disease. So we are living precariously at the present. Many scientific bodies say that the earth's carrying capacity is four billion people, and we are well past that now by a wide margin. We are heading for seven billion! We don't have enough data or experience with genetically modified plants to know if they will be the basis for the next century's green revolution. Many claims are made regarding their resistance to predators and rot, time will tell.

Some of the original seeds may be lost, and, as we return to a harsher climate, we may be in need of hardy, insect-resistant crops. The yields will be lower, and much of our food surpluses may disappear in a year or two. Then what? Then we will find food prices rising. As a result, the developing countries will be hungry first. Later many in our society will not be able to afford a balanced diet. This may result in two classes of people, those with proper nutrition raising children with a better chance of growing up healthy and those whose children will be sickly. This can really polarize America.

We have been hearing about frogs and bees dying in large numbers. Some reports say that more that 50 percent of the American bees died because of parasites, weather, diseases, and insecticides. Most fruits require bee pollination to produce a crop. This also applies to most of our vegetables. Without bees, we might as well get small feather dusters to do the job ourselves.

The use of food as a warfare agent was most evident in Holland during the Second World War. Dutch farmers planted specialty crops, flowers, and the like before the war. Staples were imported. The Germans did not supply food for the Dutch, and as a result, there was widespread starvation. Changing over from tulips to potatoes did not happen overnight, and the price was paid in human lives. Any country not self-sufficient in food is a potential victim of food wars.

Then there is another possible cause for food shortages in those countries where most of the food is grown by large commercial growers. These growers may withhold food from the market to force a rise in prices to increase their profits. Large growers have inordinate clout over any government, so you will not see any punitive action taken against them.

Day One

Once our wheat and corn reserves drop to the point that we have to curtail some exports, the day is near. Another indication is one like the current situation in North Korea, where mass starvation is reported while the North Korean government is engaged in saber rattling to take the population's mind off their growling stomachs.

We have had several examples since the green revolution. These were:

- Zambia in the 1970s had a mold called *fusarium* that struck the hybrid corn strains while traditional corn was unaffected by it.
- The Philippines from 1970 to 1972 had a rice virus devastating the new green revolution rice fields. The Tungro virus reached epidemic proportions there.
- Indonesia in 1974 and 1975 faced destruction from a viral disease of half a million acres of rice planted with the new variety of rice.

These examples happened at a time when U.S. food warehouses were full, but now it is a different situation. In 1996, weather damage to crops, critically low grain reserves, an outbreak of karnal bunt, rising prices, and a hysterical futures market have raised fears of worldwide food shortages in the coming years.

Karnal bunt, an exotic fungal disease, was the reason for a declaration of an extraordinary emergency by the U.S. Department of Agriculture. Canada banned the import of all durum wheat from the U.S. and imports of other wheat and triticale were prohibited from Arizona, California, New Mexico, and Texas. According to the USDA, this represents a threat to U.S. wheat crops. One-third of the world's wheat exports come from the United States, and now those reserves are down to a three-month level.

What to do:

- Have food reserves!!
- Obtain seeds, and store them in an inert atmosphere (cans filled with nitrogen or carbon dioxide).
- You may want to invest in food-related stocks or even in commodities.
- Be sure to have vitamin supplements.
- Read about food-preservation techniques. Have supplies on hand to do this.

Day Two

Our government claims a force-majeure clause to halt grain exports. If other major food exporters like Canada, Australia, Argentina, and New Zealand follow the same course, this will set into motion a food-war scenario. Initially the countries affected by the curtailment of food shipments will try rationing food and attempt to find alternate sources. With prices rising, marginal lands will be brought into production, providing additional food.

Another start to a food-war scenario could be a curtailment of fertilizers or fuel for commercial farms. Lack of fertilizers will negate most improvements gained by hybrid seeds and other high-yield strains.

What to do:

- Depending on the causes of the curtailment of food shipments, it may be time to look at becoming a farmer or forming a cooperative that contracts with farmers for food. There are many such co-ops in existence, most of which were formed to obtain organically grown food. The way it works is quite simple: The co-op members put up the money for the seeds, the farming supplies required, and some money for the farmers. When the crop is harvested, they take the produce in proportion to their investment. Some even work on the farm at harvest time.
- You can increase the yield from your farm if you have one worker per one-and-nine-quarter acres of land. This works in spite of current mechanized farm thinking, and where it is practiced, farm output doubles, and sometimes triples.
- Practice your food-preservation techniques. Preserve food by drying, canning, glazing, salting, and other methods.

Day Three

If the shortages continue, the countries affected by food shortages will experience mass migrations, and border conflicts will erupt. This warfare will destroy much of the growing crop in the countries affected. The extensive use of land mines kills and maims the peasants, keeping those who are left out of the fields.

The U.S. government will introduce a food tax based on the nutritional value of the food we eat. This could mean a tax on steak to compensate for the fact that each pound of steak equals sixteen pounds of grain.

Our government will have to wrestle with some *unthinkable* questions:

- How much food will we keep for ourselves?
- How much are we willing to reduce our standard of living to feed others? For example, would we stop eating beef if the grains so released would feed another five million people? Or would we rather cook hamburgers and let them starve?
- Are we going to give away our food or sell it?
- Who gets how much and why?

Coastal nations will turn to increased fishing activities. Given the overfishing today, we will see conflicts between the United Kingdom and Iceland, Canada and Spain, Canada and America, Japan and Russia, Japan and China, Malaysia and Singapore, Indonesia and the Philippines, ad infinitum. This will be a replay of earlier wars over resources, but this time the commodity is one we can't do without—food.

What to do:

- Plant your garden.
- Form a fish-farming cooperative.
- Lay in supplies associated with farming and food preservation. These can range from fertilizer through salt to mason jars.

Day Four

The major food-exporting countries will be charged with genocide, and the UN will undertake studies to fix the blame. By the time this happens, the UN may have troops assigned to it from the nations facing starvation. This can be the first step in a war between the haves and the have-nots. Such a war will see huge casualties in the have-not nations, but this will not be enough to reduce the population to the point where existing crop production can feed them. The famine and attendant starvation will continue.

A substantially cooler climate could add new and powerful countries to the list of major importers and reduce Canada's exportable surplus. Among the most immediate effects would be rapid increases in the price of food in almost all countries, which would create internal dislocations and discontent. The poor, within countries and as national entities, would be hardest hit. What is happening now to the poor in India and in drought-stricken Africa could be a pale sample of what the food-deficit areas might then experience.

In many LDCs, the death rate from malnutrition and related diseases would rise and population growth would slow down or cease. Elsewhere, there might be waves of migration of the hungry toward areas thought to have enough food. The outlook then would be for more political and economic instability in most poor countries as well as for a growing lack of confidence in leaders unable to solve so basic a problem as providing food.

For the richer countries, the impact would be mitigated, at least, by their wealth. While standards of living in countries needing to import large quantities of food would probably decline, there would be little danger of starvation. Nevertheless, there would be varying degrees of economic dislocation and political dissatisfaction, the results of which are very difficult to forecast.

What to do:

- To travel overseas as an American may be hazardous to your health. If you must travel, remove lapel pins from your clothing, carry two wallets as explained earlier under the terrorism scenarios, and try not too look too much like an American. You may want to use Canadian or Mexican airports as a jump-off point for your overseas journeys.

- Invest in farmland or areas suitable for farming. Even if you don't use it, you can always make sharecropper arrangements.

Day Five

Now that there is starvation in some countries, large-scale wars will break out. Starving people are desperate and will act in irrational ways. The U.S. may start to ration food in order to feed the starving people in other countries. There will be pressure to stop subsidizing food to urban populations, as the farmers are not able to continue to do so. This will add another dimension to the scenario, a conflict between urban and rural dwellers.

Unless you are a fan of the TV show M*A*S*H or have been a medic under fire, you are probably not familiar with the term "triage." Triage is brutally simple. When the wounded are brought to a field medical hospital, they're split into three groups (hence *triage*). One group will live whether or not they are treated quickly; they are temporarily ignored. One group will die irrespective of treatment, at least in the judgment of the person making the choice; they are left to perish. The third group may live if treated at once; it is this group that receives immediate life-giving support.

Within the foreseeable future, the U.S. will be in the position of making this judgment for a large part of the world's population. The cooling trend and the consequent climatic changes pose a threat to every man, woman, and child in the world.

The nation will have to decide what criteria will be used to sell or give away food. The decision will take into consideration questions such as:

- Should we reward countries who institute strict birth-control measures?
- Can we dictate to people who have long regarded children as their only form of old-age security?
- Should we punish nations who went nuclear at the expense of agriculture?

- Can we really let individual people starve when it is their governments that have followed misguided policies?

The answers will give the U.S. an extraordinary control over the world's nations. Nuclear blackmail is a possible risk facing us.

What to do:

- Have some members of your family or friends join the National Guard or the Army Reserve.

Day Six

The increased warfare around the globe will temporarily add to the national incomes of the governments engaged in supplying armaments to the countries at war.

This is when we may see emergence of food empires. The U.S. government may decide that since most of the world hates America, we might as well flex the food muscle. Given historical precedents, such an empire would go through the following phases:

- Build an empire. Keep its people in relative servitude by capitalizing on existing feudal institutions. For example, keep India's moneylenders in operation—that keeps the natives quiet.
- Exploit the resources of each part of the empire, but keep modern technology and manufacturing at home.
- As the empire begins to break up, prolong its existence by discouraging investment in the resources that might unify the country, modern transportation and communications systems, and in areas designed to provide self-sufficiency, such as education and agriculture.
- Once independence has been granted, make long-term planning virtually impossible by letting basic commodity prices (tin, copper, iron ore, tea, coffee—these countries' major sources of foreign funds) fluctuate with world demand based on industrial economics.
- Make sure that foreign aid is highly selective so that it rewards our friends, despite their repressive measures (after all, it takes

strong measures to be absolutely reliable) and punishes our foes.

♦ When aid is given, don't aim to make agriculture self-sufficient. Instead, make sure that it goes into high-profile, high-technology areas. This ensures that the aid is spent in the donor country and maintains continued dependence on the donor for technical assistance.

♦ Assist developing countries by keeping grain prices at artificially low levels to keep the urban poor quiet. This has two predictable consequences: By destroying the prospect of a fair market price for agriculture, it destroys the peasant farmer's incentive, and it encourages absentee landowners either to produce inefficiently or to withhold their land from cultivation altogether.

♦ Having assisted in the destruction of any economic basis for agricultural improvement, lecture sternly about birth control, disregarding the fact that high infant mortality exists and children are the providers for their parents in old age.

What to do:

▪ Obtain equipment to preserve food. You may even want to go into this business on a commercial basis.

Day Seven

We will see the start of truly massive starvation deaths in many parts of the world. Countries in Africa, Asia, and parts of Central and South America will probably suffer the most. The desperate people will try to migrate to lands where food is more plentiful. The developed countries will close their borders to these people.

These migrations will have a domino effect. The pressure of refugees on Mexico's borders will probably result in a northward migration by Mexicans, that is, toward the U.S. Similarly, starving people of Afghanistan will move to Iran and Pakistan, forcing people from these areas to move to India and Iraq.

K. Eric Drexler, in his book *Nanosystems* (John Wiler and Sons, 1992), argues that "using typical organic feedstocks, and assuming oxidation of surplus hydrogen, reasonably efficient molecular manufacturing processes are net energy producers" (p. 433). In other words, "a molecular manufacturing process can be driven by the chemical energy content of the feedstock materials, producing electrical energy as a by-product" (p. 428-9). Whether this technology can be adopted in time with its promise of food, energy, and manufacturing is not clear today. If it is adopted, the food crisis will be over for the users.

What to do:

- Maintain a radio watch to be aware of happenings around the globe. You may find that due to starvation in other lands, many of our imported commodities will be in short supply. You can make money by investing in domestic suppliers of these commodities.

Day Eight

The migration of starving people will focus on countries self-sufficient in food. They will crowd into anything that floats to escape the famine in their countries. This will lead to arming of all the affected frontiers. The United States may have to fortify the border with Mexico and have a standing army of five million. The Coast Guard will be enlarged to at least four times its present size.

As their crops fail, the underdeveloped nations will become even poorer. Their crying need to be fed will force them to divert investments from areas that could make them self-sufficient later to buying food reserves overseas now—if they can buy them at all. To add to the pain, many fertilizers and modern pesticides are oil based, meaning their prices will skyrocket, too.

What to do:

- If you live in a border area join the area patrol force.

- Extend your garden. There should be enough unemployed people around to sharecrop your additional area.

Day Nine

Canada and the U.S. will most likely have a joint approach, both countries being net exporters of food. Be prepared for an escalation in the world situation. Some countries with nuclear weapons may resort to blackmail to get their food supplies.

What to do:

- Make sure that you have access to a fallout shelter and have on hand the radiation-measuring instruments.

Day Ten

The situation can quickly deteriorate into a general warfare scenario. Governments facing mass starvation are not known for rational decision making.

W hat in heaven's name is political survival? It is the art of surviving the very governments we have elected. It is living in a country braced against one part of its population wreaking havoc on another part. It is surviving as a minority. And it is coping with extreme elements vying for power.

We practice this every day in every country on the face of the earth. It becomes a potential survival scenario when violence accompanies this political tug of war. The violence can range from police brutality to armed insurrection. Your goal is to avoid being a victim of this violence.

There is an old saying that some people are discovered and some people are found out. The latter applies to many of our politicians. There is something called "Washingtonitis," which turns public-spirited citizens running for office with high ideals into bottom feeders. After a couple of years in office, they are arrogant enough to say that only they can fix things. So things get worse. They can even set the people against the government. The scenario that readily comes to mind is the emergence of an oppressive government, which many say we already have.

One must always look upon with suspicion any sweeping government program that states "if it only saves one life." To carry this to its extreme, why don't we have doctors on street corners to save a potential heart attack victim? Even more to the point, the speed

7

Political Survival

limit on the interstate highways should be twenty-five miles per hour if we were to follow this philosophy. Yet when it comes to infringing on people's liberty, this "saving just one life" seems to have infected all governments of late. In the name of security and protection, our liberties are taken away!

Our political process works at a glacial pace. For example, if one pressure group succeeds in passing a law prohibiting the discharge of farm effluent in a watercourse, the losing group will ensure that the enforcement agency's budget is cut to the bone. So, even if there is a law, your chances of being caught are close to zero. This changes and becomes nightmarish for the general populace when some ideology overwhelmingly wins in an election and then the government proceeds to carry its mandate to the bitter end. Speedy changes in regulations are usually poorly thought out and result in all kinds of misery for many people. These people will react in some way, ranging from demand for new legislation to an open revolution.

One reason for the relative inefficiency of new laws being enforced is the nature of the bureaucratic animal. In its natural state, it grows in size and sphere of action until checked by an outside force. This check is usually in the form of financial constraints. A second check is in the form of other bureaucratic entities trying to enlarge their spheres of activity. This bureaucratic poaching sometimes is enough to keep the government out of private lives. If it wasn't for these limiting factors, we would all be working for the government—a scary scenario.

The U.S. Constitution and Congress established a body of law to govern the government. The government is meant to be the servant, not the master. The New World Order is trying to establish a rule of law based upon the assumption that governments exist to control the people. This control can be achieved by UN troops from China policing the U.S. and American UN soldiers policing Rwanda. Food for thought.

Day One

If a minority being persecuted overturns the government in a coup d'état, then becomes an oppressive government, then starts a religious war, you have all the makings of a full-blown political crisis. Could it happen to us? You bet. We have all the makings for it. Due to the poor economic climate for many, the downwardly mobile will become increasingly intolerant of others who are different. The difference can be lifestyle, race, religion, or whatever. In hard economic times, people who appear different are singled out and are sometimes accused of causing the hardships besetting the nation.

Many governments will foment this strife as it will focus people's attention away from the real cause of their problems to a convenient scapegoat. This is particularly successful if the minority has a reputation for sharp dealing or avarice.

Riots are part and parcel of political problems. People caught up in a riot are unpredictable and will do things they would never otherwise contemplate. This unpredictability is the major danger for people caught in the area of a riot. Since arson, blackouts, looting, and random shootings are part of a riot, the police reaction is likely to be swift and violent. Do not make yourself a target during these times.

What to do:

- Keep informed about national and regional trends. Be aware of movements singling out groups of people for economic and other problems.
- Be circumspect when regulations requiring the registration of firearms and similar devices are enacted. Cache most of your firearms and ammunition.
- Always view with suspicion any government regulations requiring registration or payment of taxes on anything you already own. Many times this is only a ruse to find out who has what.

- As protest increases, the authorities will make the ownership of gas masks illegal. You will be required to turn in any you have in your possession. Avoid doing so.

- If you are caught in a riot, try to move with the crowd, but at an angle to get away from the pack. If you are home, stay there unless your house is on fire. Be especially careful if you are caught in a race riot. If you are the wrong color, get away! Hijack a car if you have to, but get away. At home keep away from windows.

Day Two

By this time you will have an idea which of the following scenarios apply. Turn to that section. Always treat new regulations with a healthy dose of skepticism. Look behind the face of them to discern their true intent.

Coping with an Oppressive Government

In order to combat terrorism and crime, many federal and state governments are advocating the use of identity cards, compulsory address-change registration, firearms registration, and confiscation of certain classes of firearms. Increasingly, governments are using oppressive techniques on the general population in order to catch a few terrorists or criminals. This starts a futile process. More and more controls will be needed to know who is doing what. This intrusion into the family home is alien to the North American psyche.

Just saying *no* to the tax collectors will become harder. A misguided and desperate government will send out a message to take more from the rich to waste on government programs that don't work. "Data rape" will be the name of the game, as tax collectors and other bureaucrats stick their dirty little hands into every aspect of your life. Your financial privacy will be violated. The most successful people will be victimized the most, and for all you have to pay in taxes, you will get less in terms of good schools for your children, retirement benefits, and so on. These will cause slogans like "UP THE IRS" to appear on buildings, thus further adding to the government's paranoia.

You would be shocked to know how little privacy you have left. Introduction of new technology reduces it even more. Imagine a "smart card" that would not let you make a grocery purchase if you are more than two miles from your home. The technology is here today to enable the government to do just that. If you looked into your file at the FBI, the state police, the motor vehicles office, or your credit bureau, you might feel like a rape victim.

This information, coupled with the government's desire to be Big Brother, is the biggest current threat facing us.

George Washington said: "Firearms stand next in importance to the Constitution itself. They are the American people's liberty teeth and keystone under independence. To secure peace, security, and happiness, the rifle and the pistol are equally indispensable. The very atmosphere of firearms everywhere restrains evil interference—they deserve a place of honor with all that is good." Noah Webster said: "Before a standing army or a tyrannical government can rule, the people must be disarmed—as they are in almost every kingdom in Europe. The supreme power in America cannot enforce unjust laws by the sword; because the whole body of the people are armed, and constitute a force superior to any band of regular troops that can be, on any pretense, raised in the United States." From that we went to Attorney General Janet Reno, who said: "Waiting periods are only a step. Registration is only a step. The prohibition of private firearms is the goal."

If you have ever contributed to any cause or organization not popular with a budding dictatorship, your name will be in some data bank maintained by the government. How is that possible? Easy. Direct mail companies keep extensive records and these can be seized by the government.

Governments sometimes hide security agencies in other administrative departments. In the United States, the Secret Service is part of the U.S. Treasury. In other countries, favorite departments are census, press agencies, mapping, post office, customs, immigration, and the all-time favorite, the internal revenue services.

If you don't believe that oppressive government can be harmful to your health, consider that in the twentieth century alone close to 200 million men, women, and children have been shot, beaten, knifed, clubbed, tortured, burned, frozen, crushed, worked to death, buried alive, blown up, drowned, bombed, or killed in other ways by their own governments. A bad government can be as deadly as the Black Plague.

Day One

When address-change registration is compulsory and identity cards are issued, Day One has either arrived or is very close. This may start out benignly. Americans will be required to register at the post offices. This will apply to all residents and visitors to the U.S.

Once you are registered, the government may want to keep an eye on you. This may be in the form of a local registration card or some type of federal identification.

What to do:
- Incorporate one or more companies.
- Arrange for your phone, electricity, gas, oil, and other services to be billed to your corporation.
- Ensure that your corporate address and your home address are different. This can be as simple as having a post office box for your corporate address.
- If you have several companies, ensure that the ownership is not the same for all of them.
- You should cache firearms and ammunition in several places. A cache is different from hiding in that the cache is for long-term storage.

Day Two

Day Two will feature a need for permits to visit border areas and a security check on those living in the border regions. Based on the communist doctrine, the border region is usually a ten-mile (sixteen-kilometer) strip. Comprehensive registration of all persons is an initial step for a potentially oppressive government to take; first they must know who is living in their jurisdiction.

The government will set up committees in schools, workplaces, and apartment buildings to "guide" citizens. There will be snitch lines to turn in those who are opposed to the government. These snitch lines will be irritating in the beginning, but later on they may be a ticket to labor camps.

What to do:

- At this point, you should have a second set of identity documents. This should have been done at Day One, but it is still not too late.
- Have a cutout. That is to say, your bills and the address on your documents should be at a place where you keep a set of clothes, but do not actually live.
- Beware of any new acquaintances. The government is now collecting data on potential opposition to their plans.
- The first rule of intrigue is not to be stampeded into any unusual act—especially if you are being watched.
- The second rule of intrigue is not to give the appearance of wrongdoing—especially if you are not (at that time) doing anything wrong.

Day Three

By now the government feels that it has partial control over its citizens. The newly issued identity cards will be required for all transactions. Data banks are linked so that if you skip a day from work on the pretext of sickness and buy a ticket to a ballgame with your credit card, you may be in deep doo-doo. There will be a government seizure of all airports and aircraft, and a federal transportation agency will control all manners of transportation.

Using computer terminals in police cars, the authorities will carry out large-scale razzias (*Webster's*: A plundering and destructive incursion). These start with large police units cordoning off urban blocks. Once cordoned off, everyone within it must show the new identity documents. The police will check through computer terminals whether these are valid. Those with false documents will be detained. A complete search of all buildings will be carried out. These police units will have metal detectors and other equipment. Anyone with prohibited items will be detained.

Students in schools will have to report on activities at home. It is likely that special "counselors" will be added to the schools' staffs.

What to do:

- Use your second set of identity papers to purchase items you want to set aside.
- Start bartering for those items that the government controls.
- Get into the manufacture of improvised weapons to obtain other controlled supplies, such as ammunition.
- Arrange all traceable transactions on the basis that you buy a little extra each time you shop. This will not alert the controllers. They may think that you are overeating. You are establishing a pattern of purchasing a little extra.
- You can use the extra supplies at your real address as they are needed or trade them off for other items.
- Paint the wrong house number on your residence. Try to use a nonexistent number. This will confuse authorities and may even make a search warrant null and void (if search warrants are still required).

Day Four

The government, in the name of crime control, will institute computer indexes of *potential* criminal behavior. God save you if you are

eccentric. You may answer your phone and hear a computer-synthesized voice say, "Citizen Smith, you are in violation of the 'loud music after 11 P.M.' regulation, and your toilet flushed four times in the last hour. Stop the criminal activity."

This government interference will give rise to opposition movements, and the rise of these movements will further add to the paranoia besetting the authorities. The government at this point will do what the Soviets did, which was to keep their phone system in a mostly useless condition in order to cripple internal dissent.

What to do:

- It is prudent to find out what constitutes "criminal behavior." See if you can use your contacts to find out the criteria for index points, such as antisocial attitude, conviction for loitering, and so on. Once you accumulate enough points, they may send you to a "reeducation camp."

Day Five

To maintain control will require even more controls. You will see your remaining freedoms sucked into a black hole called "safety of the citizen." The government's desire to maintain control will extend to some very trivial areas. If not having a dog tag becomes a major offense, watch out. The government will force relocation of people from hard-to-control areas to "safe" areas. These new locations will have federal housing and control over the people.

What to do:

- Try to determine who the neighborhood informers are. If you see someone arrested and he comes back a few days later saying that the authorities made a mistake, be careful. He may have been turned and now is an informer.
- Find out where the microwave-relay towers are located. A well-placed bullet in the horn can disrupt communications. Any repair crew can be ambushed, adding to the discomfort of the dictator.

Day Six

The government will add more and more control regulations on top of those already enacted. These will be in the form of seizure of all privately held communications devices, *excessive* food supplies, and private transport vehicles, as well as complete confiscation of firearms, crossbows, longbows, swords, and bayonets.

What to do:

- You should have several caches of your supplies, so if one is found, you have not lost all your goods.
- Very circumspectly, start the manufacture of crossbows and similar items, which can be bartered for other supplies you need.

Day Seven

To build the new "safe" housing complexes, the government will establish labor camps, and anyone found subversive, unemployed, or on welfare will be sent to one. Children in these camps will be indoctrinated by their teachers and will serve as a fertile recruiting ground for future government forces. If the situation reaches this point, the government is more than likely to reach back into history to establish a new feudal society. To give you an idea of the new pecking order, take a look at the table on the opposite page.

Now that you know what is waiting for you, you must make a decision. The decision will be whether to rely on the goodness of the new order or to take chances and remain free. Remember, the serfs, once off the land of their lord, were free in a sense. However, with today's computer systems and identification methods, you will never be free.

What to do:

- Do not move to any of the so-called "safe areas." Resist by bugging out early.

- Arrange a meeting place with family members and friends should you have to bug out individually.

Ancient form of the title	New form of title
King, Emperor	Secretary-General of the UN
Princes, Archdukes	Presidents, Prime Ministers
Dukes	Ministers, CEOs of multinational companies
Counts, Earls, Marquises	Deputy Ministers, VPs of multinationals, Party chiefs
Barons	Managers of major departments and enterprises
Baronets	Senior bureaucrats
Knights	Midlevel functionaries
Free Men	Junior functionaries
Serfs, Bondmen	Citizens
Slaves	Prisoners, concentration camp inmates

Day Eight

Around this time the system will start to break down. The trivial nature of many control measures will make most people "criminals." The American psyche does not take kindly to minute control of activities. The security forces, recently enlarged, will be staffed more and more by incompetents. This is a blessing in disguise. How could you resist a really efficient oppressive government?

What to do:
- Form resistance groups with like-minded individuals.
- If you live in an area host to concentration camps, move and move quickly.

The Breakup of a Nation

The main characteristic of a peaceful society is that despite the existence of separate and distinct groups within the society, its members still think of themselves as one. Whatever quarrels the groups have with each other, the issue of national or societal unity never arises. They consider themselves part of a single entity and will subordinate their conflicts to the nation. Separatism is the most profound sign of the lack of domestic peace. To achieve national peace, the nation must provide justice and have the ability to enforce that peace. The nation must have a monopoly on violence. If and when this monopoly is breached, anarchy ensues.

America was founded by having thirteen nations combine into one. This is reflected on the national seal with the Latin phrase *E Pluribus Unum* (From Many, One). However, the thirteen states had much in common, such as language, history, and ethnic origin. The same cannot be said for other countries. Many of them have no common culture, common language, or even a common law. That is why we have so many new countries in the world.

The Civil War was an attempt at breaking up the Union. More recently, the province of Quebec came very close to separating from Canada. The October 1995 referendum in that province was decided by only 40,000 people saying "No" to separation, a very small margin in a population of close to 7,000,000. There will be another referendum, and the pundits are saying that the "Yes" side will win.

The breakup of a nation can be relatively friendly, as when Czechoslovakia split into the Czech Republic and Slovakia, or it can be major genocide, as in the Bosnian part of the former Yugoslavia. When looking at this scenario, you must keep in mind that there is no friendly divorce. The divorce happens because the two people believe that they cannot live with each other anymore. There is nothing more brutal than a divorce settlement; both parties know each other's weaknesses and points where they can be hurt.

Let us look at some possible national divorces. Italy may split along north-south lines. Transylvania wants to split from Romania to join Hungary. The Kurds want their homeland. Belgium may split into two countries, one Flemish and the other Walloon. The Tamil Tigers want to separate from Sri Lanka. And closer to home, Quebec wants to separate from Canada. The emergence of strong central governments is a relatively recent innovation.

What about the United States? Let us look at a few states that expressed interest in the past of going it alone. Vermont, Nevada, and Montana are the outspoken examples. However, there are others who under certain conditions may want to set up a regional government and operate outside the federal system. Many states could go it alone. California is as large and as prosperous as most countries in the world.

Day One

We have to look no further than Quebec in Canada. Even though the French were beaten on the Plains of Abraham, the British victors allowed them to keep their Napoleonic Code laws, language, and religious institutions. During the 1960s, many ideologist elements started to persuade the population at large that they should have an independent country, as they were unfairly dealt with by English Canada. Poets, singers, and artists were at the forefront of that movement. Eventually, two referendums later, Canada is not asking the question of *whether* Quebec will separate, but rather *when* Quebec will separate.

To recognize Day One in this scenario is more difficult than in any other. We always have groups with us who want to be the lords of their own manure heaps. Therefore, every time the federal government undertakes a new venture, there are states that want out of it. This in itself does not lead to separation, but when this is compounded by ethnic, religious, language, and other differences, the resulting brew may raise the prospect of going it alone.

What to do:

- Have your passport up to date, and renew it one to two years before it expires.
- If you expect separation, have bank accounts in other countries.
- You can make good money by contacting individuals and companies in the soon-to-be independent country and offer them representation in the mother country.
- Always have on hand resources to wait out any kind of disturbance, be that political, religious, or any other kind.

Day Two

Look at the hen as the wisest of all animals. She cackles only after laying the egg. The newly separated nation celebrates, then the problems begin. The mother country will have its troubles, too. These troubles will start with the division of national assets and liabilities. How are they going to be split? On the basis of population or where they are located? In the case of Quebec, much of Canada's industrial milk production, poultry operations, and magnesium production is located there. It is very much doubtful that after a separation, Canada would continue to shield the Quebec farmers with marketing boards from the lower-priced U.S. products available.

Initially, the mother country will make all kinds of gestures to demonstrate to the population desiring to separate that the old country can be made to work. This will be deemed to be too little, too late by the separatists. Then there is the risk of even further separation. Many will say that if the mother country is divisible, the new country is divisible, too. This may result in cities or whole areas deciding to stay with the mother country or even to form their own countries.

Separation is not the most rational process. The economic fallout is rarely calculated, so economic hardship is experienced almost immediately. These can range from inflation to shortages.

What to do:

- Hang onto your old passport at all costs.
- Keep informed concerning the plans of the new nation.

Day Three

The troubles begin. A new nation will have to establish credit with international bankers, leaving very little money left over for welfare, government grants and subsidies, environmental safeguards, and other social programs. To keep the public's support, the new government will blame the mother country for all ills besetting the new country.

The new country will try to establish self-sufficiency, also called "autarky." Limiting imports will be the first step in this direction.

What to do:

- Make sure to have monetary reserves in stable foreign currencies. You should also have traveler's checks in a number of stable currencies.
- Be as self-sufficient as you can be. This means that you should have a skill in demand and hard reserves.
- Obtain parts for foreign cars, appliances, and tools. These will make excellent barter goods for an uncertain future.

Day Four

The issue of new passports and identity documents starts. The new money, initially pegged at par with the mother country's currency, will lose value and its exchange value will plummet. The first burst of euphoria is turning to gloom and the search for the enemies of the new state will begin. Heading the list of these "enemies" will be outspoken opponents of the separation from the old country.

There will be additional pressures for self-sufficiency by the new country. These will result in rising prices and shortages of imported goods.

What to do:

- Keep your old passport by claiming that you lost it or whatever is appropriate. However, get the new identity papers as well.
- Use the new money, but do not exchange your foreign currency assets for it.

Day Five

There will be a mass migration of people who want to be in the mother country. The emigration will undermine property values, and this in turn will add to the number of properties available for sale. The new government may turn to recruiting mercenaries to beef up the newly formed armed forces.

The government will require the exchange of all foreign currencies, including that of the mother land. Holding foreign currency will become a criminal offense.

To increase the pace of the drive for autarky, the new nation will attempt a North Korea-style self-sufficiency. This will be achieved at the cost of low international trade (mostly raw materials) and the elimination of specialization of the workforce. This loss of specialization will result in lower-quality goods and elimination in the economics of scale. We may even see a return to bread being baked in a million households rather than at centralized facilities.

What to do:

- At this point, the only real estate investment you should make is in your residence.

Day Six

To foster a feeling of national unity, the new government will make proclamations regarding minority rights and peaceful coexistence. It won't work! After all, the separation happened to assert the rights of a minority. Why would they give up their newly achieved superiority?

The self-sufficiency may lead to production of "one-offs" and limited production runs, resulting in a loss in the benefits of the existing trade regime and lower living standards.

What to do:
- If you belong to the minority in the new country, move. Leave your house if you have to. Houses can be replaced.
- Cache your foreign currency holdings, but whatever you do, do not keep them in a safety deposit box.

Day Seven

Now is the time to see whether the mother country recognizes the newly independent country. The specter of a civil war may loom. This is a very likely occurrence if the newly independent country takes repressive measures against its minorities. As economic problems and consequent unemployment increase, the new government will blame the mother country and its minorities for the troubles.

Many of the idealists behind the original separatist movement will be disillusioned. With them out, there will be room in the government for newcomers. The hardships will increase with the loss of the benefits of a larger economy's division of labor. The independence movement will look to many like a pipe dream.

What to do:
- If you live in the border area between the two states, you can earn money by operating a customs warehouse or similar facility.
- There is still time to play a major part in the new government. If you support the separation, you can rise fast and far in the new country.

Day Eight

If the new country is recognized, it will have to focus on its economy. However, if the new country is not recognized, you are facing a civil

war and all it entails. Some parts of the new country will start separatist movements of their own, leading to enlargement of the police and military forces.

What to do:
- Maintain contact in both states, and be aware of public sentiments.
- Prepare for major upheaval by increasing your supplies.

Day Nine

Civil war breaks out. There is nothing more brutal than brother fighting brother. If this happens, you must take a position. This is one scenario you cannot sit out. Sitting on a fence is uncomfortable at most times, but in this scenario it can be deadly.

What to do:
- Regardless of the side you have chosen, you should join the military reserves or an auxiliary security service.

Day Ten

We are now either in a warfare scenario or watching two countries trying to get their acts together. We may even see further splintering of the new country into religious, ethnic, linguistic, and other states. This may even go as far as designer communities. The way things are going, this may be the trend of the future.

Minority Group Survival

There is a tendency in the human psyche to blame others when tough times arrive. The Germans blamed the Jews for their recession. In the aftermath of 9/11, some Americans blamed anyonse who appeared (correctly or not) to be of Middle Eastern descent. These are just two examples. Being a visible minority can be hazardous to your health in most scenarios. At best you may end up as a slave in some enclave not yet blessed by the ownership of a tractor; at worst you can be very, very dead.

Because of labor shortages, minority populations are sometimes encouraged to immigrate to a country. But as in Germany, once hard times come, these groups are pressed to return to their homelands. If they do not leave willingly, the host country will enact restrictive citizenship laws and offer free transportation back to their country of origin. The unemployed in the host country will blame the minority for their problems and attack areas where the minority lives and works.

For the minority groups, the suggestion to move to a ghetto of like people can have a twofold effect. While it provides seeming security, it also identifies where you can be found should the crowd's mentality turn to lynching or like activities. Therefore, let us focus on practical ways of coping with the majority when they start to blame you, the minority, for their misfortunes.

Some minorities at different points in history came through wars, upheavals, pogroms, and the like in relatively good shape. When a minority does, it is generally because of three reasons: one, it is not tied to the land; two, its members avoid exhibiting their wealth; and three and perhaps most importantly, even if easily identified, they present a nonthreatening posture.

Day One

When you read that a certain group is singled out for some national ill, take heed. That is usually the first step in stirring up hatred against a minority. This can follow from the preceding scenario, "The Breakup of a Nation." A new country is having troubles and will make a minority the scapegoat for its ills. It has happened before, it is happening now, and it will happen in the future.

What to do:

- If you are a member of any minority under attack, this is the time to decide whether to stay put or move. Remember, most of this continent was settled by people who switched countries rather than staying and fighting. You may have to continue the tradition.
- Have a bug-out kit.

Day Two

Around this time, thugs will attack business premises and homes owned by the minority group. The more prominent you are, the more likely that some unpleasant incident will happen to you or your family.

The government will call for cooling of the tempers while it is evaluating whether to sacrifice the minority as a scapegoat for the national ills.

What to do:

- Avoid any travel through areas controlled by those who are against you. At best you only get raped or beaten up, at worst you end up dead.
- If you are in the minority group, have a security force escort your people when shopping or doing other activities.

Day Three

There will be raids on minority areas. In the beginning, these will be unorganized, but as time goes by, you will see an organized pattern emerging. The minority group under attack will arm itself. The government will send in federal police forces to protect the minority under attack. Depending on how widespread the national ill for which the minority is blamed is, these police forces may or may not be effective.

What to do:

- If you belong to a visible, easily identified minority, you may want some members who are not easily recognizable to do the shopping and other essential functions. During World War II, blond-haired, blue-eyed Jews saved a lot of their people.

- If you sympathize with the minority group, you can be a tremendous asset in protecting them. You can provide safe houses, transportation, employment, and supplies of all kinds.

Day Four

Large-scale conflicts will emerge, followed by mass migration of the minority under attack. Depending on the central government's attitude toward the minority, we will see a genocide or a controlled situation. If the minority has a home area, it will be swollen with refugees and turned into an armed camp.

The disruption caused by the conflict will result in rising prices and shortages of supplies. People will demand that the government do something about them. The government may blame one minority or the other for the troubles. This will polarize public opinion against the targeted minority.

What to do:

- Establish an information network to be in position to evaluate what is likely to happen.
- If you have extra supplies, you can sell or barter at a profit.

Day Five

The government will step in to protect the minority under attack, first with increased police presence in the area where the minority group lives and works. If this does not control the violence, a curfew will be put into effect, and troops will be brought into the contested area.

If the minority has an identifiable group preying on it, it will retaliate. An ever-increasing cycle of violence may develop at this time.

What to do:

- Supplying the minority under siege may be a risky, but profitable enterprise.

Day Six

The central government may take one or more of several possible steps. These can range from establishing protected reservations for the minority under attack to indifference to their plight. Whatever action is taken, the minority will realize that they are on their own.

What to do:
- Depending on your attitude toward the minority under attack, you can do several things. You can support them by hiding them, providing food and other supplies to them, and even arming them.
- If you are part of the minority, your choices are limited. You might as well get to like a reservation existence.
- If you belong to a minority not under attack, you may want to think about how to cope when your turn comes.

Day Seven

The minority will be conspicuous by its absence on the urban landscape. Countries from which the minority originated will provide covert aid to those living in the United States.

What to do:
- If you belong to the minority, it will be prudent to make plans to move to a country where you are the majority.

Day Eight

To reduce violence and placate citizens, the government will undertake large-scale resettlement of the minorities. All this will be done in the name of protecting them. Based upon what has happened to North American Indians, we will probably see them moved to marginal lands and forgotten.

Appropriate Behavior During a Coup d'État

No, this is not intended as a chapter on etiquette, in spite of the title. When there is an armed takeover of a government, it is very difficult for the players to separate friends from enemies. Many innocent bystanders pay with their lives for that confusion. This confusion can help you to obtain supplies—and I do not mean looting! Why get into a firefight over a couple bottles of liquor when you can drive up to a warehouse and drive away with a truckload? The difference between foraging and looting is the same as between making love and a gang rape.

The technique of paralyzing and seizing the technological power points of the state is the tool of a coup d'état. It is a relatively modern practice coming into favor with the evolution of professional bureaucrats and standing armies. The modern state controls people through the data it has on them and through allocation of resources. The bureaucrats are the ones who make these happen.

We all know that many civil service jobs are given to political associates rather than on the basis of professional competence. If the bureaucrats are linked to the leadership, you must have a palace revolution for an illegal seizure of power. If the bureaucrats are not linked to the leadership, an outside coup d'état is more feasible.

In many countries, you will find ethnic bonds in the senior levels of the government. Look at Africa where rulers appoint members of their tribes to key positions in the governments and security services. Bureaucracies, to function in an efficient manner, have a chain of command and standard procedures to follow. When an order comes from the appropriate source, it is followed by a stereotyped response. *This is the key to a successful coup d'état.* During the coup, the use of parts of the state organization will ensure the "machine" will carry out the new orders. If the machine is very sophisticated, a coup d'état is very difficult to pull off.

Other names and forms of a coup d'état are revolution, civil war, pronunciamento, putsch, liberation, war of national liberation, and insurgency. They differ only in the level of violence. The ultimate aim remains the same, to seize the levers of power for controlling the state. Since this is not a textbook, let us look at the hows rather than the whys.

If you intend to have your own coup d'état, remember to use the standard procedures. These are:

- No information is to be communicated except verbally.
- No information is to be communicated except on a need-to-know basis.
- All communication links from inner to affiliated members are to be on a one-way basis.
- No activity is to be carried out by an inner member if an outer member can do the job.

A good coup d'état should be bloodless. However, the best-laid plans have glitches, and as such, your life may be in danger.

Day One

If your government cancels elections and rules by force, using the military for domestic police duties, you can anticipate that some group or groups will have a desire to change this situation by force of their own. You may see frequent razzia raids. These are a European type of police measure. As explained earlier, the police arrive in force and seal off several blocks. Everyone must produce identity cards to leave the area. The police also search each building in the cordoned-off area. When these things happen, you can be fairly sure that Day One has arrived.

In a coup d'état, the defenses of the state must be neutralized. What does this mean in a democratic state like the U.S. or Canada? In America, the armed forces, FBI, and other security organizations,

along with their intelligence-gathering arms, must be brought under control. Similarly in Canada, the armed forces, the Royal Canadian Mounted Police, and security agencies must be controlled in order to have a successful coup d'état. A tall order, however, due to a polarization in our society, it may come to pass.

When you see a large number of military vehicles in the major cities, in particular the capital cities, a coup d'état may be underway or is being snuffed out. Usually the first twelve to twenty-four hours are critical. If the coup does not succeed in that period, then the state will have time to bring in other military units from outlying areas. Then it is over or you have a civil war to contend with.

What to do:
- Once again, it is handy to have two sets of identity documents.
- As always, have a bug-out kit.

Day Two

This is the day of the coup. Stay put, and avoid firefights. Remember, you want to survive, not conquer. This will not be a fight between determined civilians and the military of the state. The modern infantryman has so much firepower at his disposal that no civilians can overcome the armed forces. We may see one part of the armed forces squaring off against another. In a realistic scenario, we may find that a minority group is overrepresented in the military and security forces of the state. This may be because the military provides the only way out of poverty and marginalized existence. This may be why the regular army has a higher-than-normal proportion of African Americans in its ranks. In contrast, the National Guard is predominantly white. Should the armed forces become politicized, a fertile ground for recruitment would develop.

There are certain symbolic targets, such as the White House. The seizure of symbolic targets is very important to the leaders of a coup.

What to do:

- Keep a low profile, and listen to the news. Use your scanner radio to pick up international shortwave broadcasts on how outsiders interpret the ongoing coup.

- Stay away from windows. You may draw fire from one of the participants.

- If you are a well-known supporter of the government being overthrown, move and move fast to another area of the country or out of the country altogether.

- Avoid communication centers, military installations, government offices, banks, police stations, and the like. Fighting could be very heavy around those areas.

- In the confusion of the coup d'état, you may have an excellent chance of removing police files pertaining to you and your friends. This is a major opportunity to clean up your past, particularly if you were a police informer for the administration under attack.

- Be sure that you have filled all available containers with water. Utilities are likely to be interrupted for the duration.

- Roadblocks in sensitive areas of the capital and control of the major airports are stock in trade during a coup d'état. Avoid traveling if possible. If you must travel, don't go near the capital city.

Day Three

The targets have been seized, the loyalist forces isolated, and the rest of the bureaucracy neutralized. This marks the end of the active phase of the coup. However, the situation will be fluid, a counter-coup is very possible, as well as a fight for power among the coup leaders. A new provisional government will start to issue new identity cards. Do not trade in your old ones. Claim that they were taken away from you during the coup d'état.

To freeze the situation, a total curfew will be established. This will include a shutdown of all forms of public transportation, closing

THE SURVIVALIST'S HANDBOOK

of public buildings, and an interruption of all telecommunications services.

What to do:

- Register for the new identity documents. Register early and register often. It is likely that in the initial confusion, you can get more than one set of identity documents.
- Hide your old identity papers as well as any material that would identify you as a supporter of the previous government. You may need them later if the pendulum swings back.

Day Four

Now the government will start a systematic roundup of the supporters of the previous regime. If you are such a person, you should have moved abroad on Day Two. Many will be jailed to settle old scores, even those who were apolitical.

There will be a massive information campaign in the mass media to reach the population not involved in the coup. The authorities will make a tacit deal with the bureaucrats and soldiers, assuring them that their careers will not be threatened.

What to do:

- If you are still at liberty, your chances of staying free are excellent. This is the time to figure out how to make a living under the new administration.
- Listen carefully to what people are saying. This may give you an early warning of a coming countercoup.

Day Five

Governing begins in earnest, starting with proclamations stating that we should all work together, followed by regulations to exclude the supporters of the previous regime from participating in the government. Children of the supporters of the fallen regime may be deprived of university education.

This period will provide wonderful opportunities to those supplying crowns, flags, decorations, royal palaces, mistresses, country hideaways, and other necessities of state.

What to do:

- If you are going to stay in the country, you might as well make sure that you can make a comfortable living. Given the changes in preferred suppliers, this is your chance to get in on the act. The downside is that if a countercoup is successful, you will be identified as a supporter of the present rulers. You make your choices, and you have to live with them.

Day Six

The new government starts to feel secure. As a result, many of the control measures are relaxed. This is another chance to move if you do not like the new government.

What to do:

- Keep a low profile, and maintain a careful watch over what is happening in the country. By this time, you have recognized the advantages of living far away from the capital city.

Day Seven

By now you are either happy with the new regime or you are not. If not, start forming a resistance group. Depending on how the government is running the country, it may be time to get ready for another coup d'état.

What to do:

- If you are in the resistance, follow the tried organizational formula of having a cell structure for your organization.

Day Eight

The new government will make every effort to consolidate its position. This is usually done at the expense of freedoms.

The Left Versus the Right

The labels of the "left" and the "right" are rather fuzzy. What we are concerned about is the effort of powerful left-wing and right-wing movements to seize control of the state from parliamentary assemblies and thus place a narrow elite group in power. The increasing chasm between the haves and the have-nots translates itself into an increasing division between the left and the right.

What is at stake here? At stake are your freedoms, property, and even your life if things go the wrong way. The so-called left traditionally espouses causes like welfare, social justice, full employment, union activities, minority rights, and open immigration policies. The right wants to reduce government involvement in our lives and business activities. The haves mostly belong to the right while the have-nots mostly cant to the left. Over the last forty years, the U.S. has seen swings in both directions. Where the situation gets sticky is when one side is in power for too long, although time is not as significant as the pace of change the government introduces.

Let us look at the pace of change. The Clinton and George W. Bush administration introduced wholesale changes. These changes were greater than during the combined Reagan and Bush years. The result has been a polarization between the haves and the have-nots. Now we find that instead of a civilized debate, we are facing violent confrontations. The availability of knowledge for wreaking mayhem could only be lessened by a frontal lobotomy of all retiring servicemen and the removal of encyclopedias from public libraries. Not likely to happen. So we must be aware of the changes around us, both locally and nationally. We may even face standoffs between New Hampshire and Massachusetts. Think about it, in New Hampshire the license plates carry the logo "Live Free or Die," whereas in

Massachusetts there are warnings about a one-year sentence if you have an unregistered handgun in your vehicle. Talk about a different ethos for adjoining states.

And what about the problem between the California dream and the Nevada dream? The California dream is of government protecting you as you whiz along the highway in your Corvette. In contrast, the Nevada dream is a self-sufficient individual rebuilding his Super-Turtle pickup truck. As long as the economy chugs along, these differences can be accommodated, but when hard times come around, the search for a scapegoat is high on the agenda. This is the start of the great divide.

The beliefs of the left might be summed up as follows:

* All people, everywhere, want peace.
* Only their evil leaders want war.
* All cultures are of equal value. All that is needed is for us to understand them.
* Given the opportunity, any culture—no matter how retarded or vicious—will emerge into a true civilization.
* There are no savage nations, just nations that, if given sufficient American money, will achieve greatness with the developed nations.

The right in general does not believe the above.

Day One

When major violent confrontations start between political groups and when religious differences get out of control, you will see the arrival of Day One. Some say that it is here now, but they are wrong. The level of violence is very small at this time. What we have is a noisy debate with flashes of occasional violence resulting in an ever-spreading pool of angst. At the present, both major parties are very much alike.

Day One can start with a debate between the pro-life (right) and pro-choice (left) factions. The pro-choicers may push through a tax amendment to increase taxes on pro-lifers to pay for taking care of children and their mothers who might have otherwise sought an abortion. The tax increase on pro-choicers would be very small, simply the cost of one or two visits to a clinic.

The above possibility would lead to big-time confrontations between the two groups. This would in all likelihood extend to welfare, farm support payments, foreign aid, membership in the United Nations, Medicare, and other issues. The battle lines will then be drawn.

What to do:

- Have your bug-out kit ready.
- Avoid arguments with your coworkers about politics. They may remember later and target you.
- Increase the quantity of ammunition on hand. A good firefight may consume as much as 500 rounds of ammunition in just one day.

Day Two

There are violent confrontations and martyrs emerge for both causes, à la "Remember the Alamo." The government will attempt to placate both sides with inaction. This will give rise to accusations by both groups of favoritism. A pragmatic government would eliminate laws, regulations, and tax breaks for special-interest groups, but this is not likely to happen. Election funds would greatly diminish if special-interest groups did not exist.

What to do:

- If you have any funny political bumper stickers, remove them now.
- Remove any political lapel pins, badges, or hats from your clothing, and get rid of political literature on the coffee table in the living room.
- Unless you are very strongly favoring one or another of the participants in the political struggle, avoid political rallies.

Day Three

After a spectacular confrontation, the government will call for reconciliation and forgiveness between the warring parties. Depending on whether the government in power is leaning to the left or to the right, there will be government-sponsored protection for that group's political rallies. This measure sets the stage for a potential civil war.

What to do:
- If you reside in an area prone to marches, rallies, or other public gatherings, you may want to move and either sell or sublet your residence for the duration.

Day Four

If the side backed by the government is losing, they will be armed by "sympathizers" in government circles. The other side will arm, too, by obtaining whatever arms can be had. This arms race will lead to more confrontations.

The federal government may temporarily halt the sale of firearms and ammunition. Sporting goods stores, gun shops, and other stores may see their stocks seized for the duration.

What to do:
- If you live in an urban area, you may want to move your family to a safe area in the country.
- If you own a gun shop, this may be the time for a small fire affecting your records.

Day Five

Small-scale armed conflicts will be the daily fare. The armed forces will be mobilized to keep the two sides apart. Depending on the level of the government's support of one of the factions in the conflict, this may mean repression for the antigovernment faction. The involvement by the authorities will signal a change in the conflict. It may even become like a civil war.

What to do:

- If you are a member of the National Guard or the reserves, you should decide whether to go on an extended medical leave or upgrade your training in urban warfare.
- Add to your supplies. Prices will rise, and availability will decrease.

Day Six

To control the violence, personal freedoms will be limited. These can range from curfews to outright bans on public meetings. The opposing sides will continue the arms race, and even museums will be looted for weapons. Police may raid the headquarters of some political parties to obtain lists of members and contributors.

What to do:

- You might as well decide which side you are on at this point and act accordingly.

Day Seven

Now one of two things will happen. The government will espouse one faction or another, taking us toward a civil war, or there will be some de facto cease-fire while the politicians work out a compromise. Turn to either the "Oppressive Government" or "Breakup of a Nation" scenario.

The Population Bomb

The planet's population has more than doubled since 1950. It is growing faster than our ability to feed it. Birth rates are very high in the developing countries. The rural poor flock to the cities and add to the urban sprawl. On the other hand, population growth in the developed nations is minuscule and in some instances declining.

The growth in population places an ever-increasing burden on the ecosystem to feed and house the new mouths added each year. Concurrently, there are warnings that the earth's carrying capacity is four billion people. In stark terms, this would mean that three billion must die. An additional problem is the impact of any disaster on the population. In the final analysis, it does not matter what kind of disaster we are talking about. If you have to evacuate people, you have twice as many to move, feed, and house than you did forty years ago. This magnifies the effect of any disaster.

Close to a billion people suffer from malnutrition and starvation today. The population of the less-developed countries accounts for close to 80 percent of the total population of the planet. This requires a huge amount of cereals to feed them, which in turn places pressure on the food-exporting countries. In America, close to one half of our cereals goes to feed cattle. To produce one pound of meat takes sixteen pounds of cereals. Given the current U.S. consumption of meat, more than a ton of grain is required to feed the average American, while in China about 400 pounds of rice is consumed by an average person. Can you imagine the impact if they switch to chicken?

The old equation of "one man, one plow, one acre" gave way to large agribusiness enterprises. These have enabled us to feed larger populations. Yet all we need is drought in the wheat- and corn-growing regions and food stocks will be at an all-time low. What this will do to food exports to less-developed countries is frightening. Reduction or elimination of food aid would only be the first step.

In the Third World, the migration from the land to the cities has resulted in huge urban slums. The poor come from the rural areas looking for work, education, and medical care. Once in the cities, they are exposed to the lifestyle of the rich through television and movies. This gives rise to potential urban violence, which forces the government to spend most of the external aid in the cities. Most of this has gone to subsidize food prices at artificially low levels to

pacify the urban poor. Very little is spent on improving agricultural productivity.

When marginal land is brought into production, in most cases the land rapidly loses its nutrients. If no imported fertilizer is available, the land will be worse off than before. The farmer giving up farming his poor soil also drifts to the cities.

Some of the largest cities are in LDCs. A few of them are listed below:

City	Estimated Population of the metropolitan area in millions
Jakarta, Indonesia	23
Delhi, India	22
Mumbai, India	22
Mexico city, Mexico	21
Manila, Philippines	20
São Paulo, Brazil	20
Shanghai, China	19
Karachi, Pakistan	16
Kolkata, India	16
Cairo, Egypt	15
Tehron, Iran	13
Lagos, Nigeria	12
Rio de Janeiro, Brazil	12
Lima, Peru	8

The list goes on and on.

The developed countries are giving away or selling food at very low prices to aid these countries. This leads to peasants abandoning farms in even greater numbers as they can't compete with the low-priced imported food. Should the food shipments falter or cease for any length of time, we would see riots in some cities that would make any war pale in comparison.

The key to feeding an ever-expanding population is ever-expanding food production. Once we reach a plateau in yield per acre, this will signal the limit to the carrying capacity of the planet. This number is anybody's guess at this time, but depending on whom you listen to, the numbers range from 4 to 35 billion people.

In the case of a worldwide disaster, the first continent to go under will be Africa. Given the existing land bridge between Africa and Asia and then onto Europe, it is no wonder that some are expecting the Middle East to be the site of Armageddon.

8

There are a few conceivable scenarios in which we may have to consider the survival of a large part of the species with little or no warning period. The scenarios we hear about range from planetary catastrophes to an invasion of aliens. These are extreme scenarios, yet preparation for them starts off with the mundane advice given for others—that is, acquire skills and supplies while the going is good. Just thinking about what could happen has some survival value. You will eliminate many actions that obviously will not work.

To give you a flavor of what can happen, read a few "what if" books of fiction, like *No Blade of Grass, Alas Babylon, Malevil, Lucifer's Hammer,* and *Omega.* Videos like *Independence Day* and *Red Dawn* will also aid in formulating ideas pertaining to "what if" scenarios. Read science fiction; today's fiction is tomorrow's fact. With a little experience, you can formulate your own potential scenarios. Some scientists believe the dinosaurs became extinct due to an asteroid strike! We could be the next species to pass from history.

Another bizarre occurrence could be a perturbation in the orbit of the moon that would affect our tides. Or the release of carbon dioxide due to sudden heating of the oceans or even a sudden change in the jet stream. Many people told us during the preparation of this book that these things will not happen, but what if they do? What will you do then?

Day One

Constant preparation for disaster scenarios can be harmful to your mental health and pocketbook. However, just because someone is paranoid does not mean that no one is after his hide. Sometimes it pays to wait out this type of scenario. Disaster affecting the whole planet is the hardest to deal with, since we are faced with the possible extinction of the human race. Yet in almost every disaster ever experienced, there have been survivors.

What to do:

- As always, your bug-out kit should be ready to go.
- Keep informed. With inexpensive multiband radios and TV coverage of world events, there is no excuse for ignorance.
- Arrange your affairs so that you can work from home.
- You should strive for self-sufficiency, which will help you to deal with any change while keeping comfortable.
- Evaluate news with the attitude of a cynic. Do not jump to conclusions.

Day Two

Turn to one of the specific cases described in the following sections. This field is too large to cover every imaginable scenario. You may want to write your own as the events unfold.

Planetary Catastrophes

Imagine the earth's axis tilting and dumping ice sheets from the Antarctic continent into the drink. Or a large meteor striking our planet, like the one that wiped out the dinosaurs. These are remote, but possible occurrences. We have evidence in volcanic rock of the shifting of the earth's axis several times in the past. So, although remote, these things have happened before.

These catastrophes impact the whole planet. If you are an accountant, lawyer, social worker, computer programmer, or bus driver, you'd better take up a few hobbies like gardening, woodworking, and machine-shop skills for you are likely to be severely underutilized.

The best way to cope with a planetary catastrophe is to be able to trade skills for the items you need. Before you can undertake trading your skills, you must have adequate resources to get through the initial period. Same advice as before—have at least three to six months' food supplies handy, water-purification equipment, protection devices, and other equipment. Should the earth's axis shift or a large meteor hit an ocean, you can expect major tsunamis (tidal waves). If you are living in low-lying areas on the seashores, islands, or other susceptible places, your ark should be ready to ride this one out. You should also have the supplies to enable you to reach a habitable portion of the globe.

In the event of the axis tilting, asteroid striking, or ice sheets slipping, anything from a cold summer to an ice age will result. Any asteroid hit will cause increased volcanic activity, followed by plenty of dust in the atmosphere. Similarly, the icebergs will cool the oceans. Most of these potential disasters will result in cooler weather.

Recent evidence indicates that the earth's wobble is increasing. This is partially blamed on the huge water reservoirs associated with hydroelectric projects. Increase in the wobble may lead to the ice cover from the Antarctic sliding into the oceans. Should that happen we shall see an increase in the water levels and most of the existing ports will be underwater. The tsunamis generated by such an event will devastate most coastlines, and some islands will disappear. The cooling of the oceans will enable the absorption of additional carbon dioxide, since it is more soluble in colder waters. This will reduce greenhouse gases and may signal the beginning of another ice age.

Should the earth's axis tilt, civilization as we know it ends for all practical purposes. A tilt will also be accompanied by tsunamis.

In addition, you will have unprecedented earthquake activity, atmospheric disturbances, and a virtual cessation of transportation networks. Widespread famine will result. Most cities have food supplies to feed the population for only two weeks. With transportation networks interrupted, the population of these cities will erupt into the countryside in search of food.

Day One

This one will most likely come without warning. One minute you are sipping a lemonade in your backyard, and the next you're looking at a wall of water ten stories high coming straight at you. This is when it is handy to have all your essential supplies packed and ready to go.

What to do:
- Establish prearranged meeting places with your family and friends. This way you can get together even if all communication networks and devices are useless.
- Be sure that you have emergency supplies and equipment packed and ready to go at a moment's notice.

Day Two

The catastrophe has happened. If you are living in a nation not affected by the catastrophe, be prepared for refugees arriving by all means of transportation. Your government will either admit refugees or seal the borders.

What to do:
- Have your family and friends move in with you or move in with them. At this time, you will need all the friends you have around you. It is easier to patrol the area when you have more people. A compact group uses fewer supplies than one spread over a large area, not to mention that a full house preempts the authorities from placing refugees with you.

- Keep up to date with events. Have a radio and satellite TV, and watch and record things as they happen. Try to figure out the extent of the catastrophe.

Day Three

If you are in a host country to the refugees, several changes in your life will be made by your government. These may include some or all of the following:

- Declaration of martial law and restrictions on movement in the country.
- People may be billeted with you.
- Rationing of food, fuel, electricity, and other essential items will be introduced. The rationing of services may be included.
- Medical personnel will be conscripted for the duration.
- If you are in an armed forces reserve, expect to be mobilized.
- The news media will be under the direct control of the authorities.
- A moratorium will be in effect on all debts, rental payments, and mortgages for an indefinite period.

What to do:

- Maintain a communications watch to learn for yourself the true extent of the catastrophe. Even no news is of value. For example, if you have the frequencies of major shortwave stations, their absence may tell you something. Your government may not tell you the whole truth.
- Sit down with your friends, and based on the information on hand, try to determine the long-term effects of the catastrophe on you and on your area.

Day Four

This is about the time you learn about the cost of the catastrophe in terms of human lives. You will also have a good idea of the

408

migration patterns of people from other places. It is quite possible that your area will be sealed off from the rest of the country.

What to do:
- Make a list of people and skills you would want in a group with you.
- Take a map, and mark the known changes on it. This will help you to understand the magnitude of the disaster.

Day Five

By now you should have an idea whether you are faced with a permanent change in our way of life. If it is a change for the long term, form a cooperative and plan how to survive under the new circumstances. These changes may involve regressing to an agrarian mode of life.

What to do:
- Organize a self-help group. In all likelihood, you will have to rely on your resources rather than government assistance.
- Do not accept newcomers unless they have something to contribute to your group.

Day Six

Democratic governments will cease to exist. This is a situation where we cannot afford democracy. The evolution of regional governments is a very likely event to prepare for.

What to do:
- Make sure that you are represented in any new form of government. You can do this by being an organizer of the new and evolving system.
- Maintain your group's independence through self-subsistence. This way you will have a chance to exercise a decision-making function.

Day Seven

This will be a difficult period. You must decide whether you will join an organized society where your freedoms are severely limited or try to make it on your own. If the loss of life was extensive, you can form your own "country." If you are in a populated area, this may not be a viable alternate.

What to do:

- Your cooperative could form the basis for a pressure group. You could find your group in a dominant position locally.

Day Eight

So you have decided to make it on your own! You will have a chance to rebuild civilization. First, you must have a large enough gene pool, otherwise inbreeding will eventually cause genetic diseases among your descendants.

The popular writer Jerry Pournelle best describes what a new world will bring: "A world with few comforts and luxuries . . . a world without rat-races . . . a world with few regulations . . . paperwork and all the other frustrations that make us psychoneurotic."

Invasion of the Aliens

Little green men from Mars? Not likely, but with all those stars out there, we must be incredibly arrogant to assume that we are the only species of intelligent beings in the universe. To get to us, the aliens must have more efficient propulsion systems than our rudimentary chemical engines. Knowing that the aliens have a higher technology base, many writers assume that they will be godlike, loving beings who will teach us.

Since we have not seen any aliens yet, we can just as well assume that they will be escapees from some intergalactic lunatic

asylum bent on enslaving the earth's population. Or we may be nothing more than dinner to them. Or we may be a fine sacrifice for the Great God Fusion or Fission. Use your imagination. Almost any nightmare is possible.

Ever since UFOs were seen by credible observers, many of us have pondered the significance of their presence. Are they scouts for an invasion, demons, casual visitors, or guardians to seal off our budding civilization? If we had the answer, we could bottle it and make a fortune. Lately the skies over Israel seem to be full of UFOs. Some psychologists say that they are an effect of electromagnetic currents on the brain. Who knows?

How does one prepare to meet the aliens? Very carefully. This is one scenario when doing nothing has its own rewards. As a cynical person would say, let others be on the dinner table.

Day One

We may not detect incoming spacecraft. At first we may think they are meteors or asteroids, but when we detect that they are braking and the speed of the object lessens, we will know that we are dealing with spaceships under intelligent control. At that point, the question will be whether they are "manned" or "unmanned." If unmanned, you can expect a flyby mission to take photographs, collect data, and relay the information to a mother ship somewhat behind the exploratory vehicle. The flyby mission will tell you nothing much about the aliens except that they are here. It will also demonstrate to you that we are not alone in the universe.

Now for the sixty-four-thousand-dollar question. Are the aliens friendly? Any race capable of traversing interstellar distances will be able to do the following:

- Knock out worldwide communication networks. Whether this will be through the use of electromagnetic pulse is of no significance. The net result will be that we will be in the same boat as if we stepped back in time to the early 1800s.
- Put out of commission all the satellites we possess.

- ◆ Control all atmospheric flights. We may see the reactivation of old battleships.
- ◆ Bombard Earth with asteroids and other space debris.

Some of these actions may be taken simply to ensure that we do not overreact when they arrive. However, if the visitors are bent on occupying the planet, these will have military significance. The problem we will face will be deciding why they are doing what they are doing.

What to do:

- If you have gotten this far in the book, you know that you should have emergency supplies on hand and the training to use them at all times.
- If you hear of large numbers of UFO sightings, pay attention. Those who saw them may not all be fruitcakes.

Day Two

They're here! Now for the question—are they tourists, refugees, occupiers, looters, real estate agents, or what? If nothing is happening, they are probably evaluating us. Wherever they land, the government of that nation will try to communicate with them. If a deal can be struck with the aliens, that nation may end up as the only remaining world power. If not, one zap and it's vaporized. Imagine if they land in Iran!

What to do:

- Leave large cities.
- Avoid travel by aircraft.
- Maintain a communications watch to find out what is happening at other locations on the globe.

Day Three

During this period, the government will attempt to communicate with the "visitors." If they have landed, expect to see the area sealed off and the residents evacuated. The security cordon will be rigidly enforced, and would-be trespassers will be shot. The potential for a mischief-maker contacting the aliens will give nightmares to the authorities. Depending on the aliens, physical appearance, rumors regarding their behavior will abound.

What to do:
- If the landing is made in your area, follow government orders and evacuate. This is one time to follow orders.
- Reduce any activities observable from the air. Until you know more, keep quiet.

Day Four

Contact is made with the aliens. The aliens are unbelievably different from what we expected based on the writings of science fiction authors. They can even look like us, but their thought processes have been fashioned in a different star system and as such even if they look like us, they will be very different. Many of our earthly taboos may seem alien to them.

What to do:
- Keep current with the news. The use of shortwave radio will enable you to hear alternate viewpoints, not just those of your government.
- Nongovernmental agencies will be on site, as well. Listen to what they have to say about the aliens.
- Remember, the aliens are different. This is one time when you must not jump to conclusions. Just because you don't like spiders does not mean that spiderlike beings are necessarily evil.

Day Five

One of two possibilities, both very unpalatable, will emerge. The first assumes that the aliens are friendly and we will be in the same boat as the Aztecs with wooden swords facing off against Cortés or the Incas against Pizarro. Neither the Aztec nor the Incan civilization survived meeting with superior technology. Most of our existing civilizations will crumble. For a more recent example, look to the cargo cults of the Pacific Ocean region.

The other possibility would be an alien invasion bent on conquering the planet. We would fight, there is no question about that. The question is: Would we win?

This marks the end of the scenarios. With any luck, you will not have to use the contents of this book. However, the way things are going, I'm afraid that one or more of them will have to be dealt with during your lifetime.

I urge you to start looking around now, evaluate what is likely to happen to you and your area, and start preparations to cope with potential threats. Look at your home and devise plans and methods for enduring the most likely threats against it.

Appendix

Suggested Publications for a Reference Library

Here is a list of publications that you should try to acquire for your library. It is not exhaustive, but should provide much of the information you will need to survive. Wherever available, the authors' and publishers' names are provided.

- *American Survival Guide*, all back copies
- *Back to Basics*, Reader's Digest
- *Barefoot Doctor's Manual*
- *Barnacle Parp's Guide to Garden & Yard Power Tools*, John J. Mettler
- *Boobytraps (Field Manual 5–31)*, U. S. Army manual
- *Breathe No Evil: A Tactical Guide to Biological and Chemical Terrorism*, Stephen Quayle and Duncan Long, Safe-Trek Publishing
- *Brown's Alcohol Motor Fuel Cookbook*, Desert Publications
- *Build Your Own Low-Cost Log Home*, Roger Hard
- *Building Underground: The Design and Construction Handbook for Earth-Sheltered Houses*, Herb Wade
- *The Canning, Freezing, Curing, and Smoking of Meat*, Wilbur F. Eastman Jr., High Country Enterprises
- *The Chemical Formulay*, CRC Handbooks (more than twenty volumes)

- *C. I. A. Improved Sabotage Devices,* Desert Publications
- *Combat Skills of the Soldier (Field Manual 21–75),* U. S. Army manual
- *The Competence Factor,* Bradford Angier, Stackpole Books
- *The Complete Survival Guide,* Mark Thiffault, editor, DBI Books
- *Designing and Building a Solar House,* Donald Watson
- *Encyclopedia of Survival,* C. M. I. C. Group, Safe-Trek Publishing
- *Engineer Field Data (Field Manual 5–34),* U. S. Army manual
- *Explosives and Demolitions (Field Manual 5–23),* U. S. Army manual
- *Farming for Self-Sufficiency,* John and Sally Seymour
- *Food Drying,* Phyllis Hobson
- *Fortunes in Formulas for Home, Farm, and Workshop,* Books Inc.
- *Gun Digest,* all back copies
- *Guns and Ammo,* all back copies
- *The Herb Book,* John Lust
- *How to Do Just About Anything,* The Reader's Digest Association (Canada) Ltd.
- *Live Off the Land in the City and Country,* Ragnar Benson, Delta Press
- *Making Your Own Motor Fuel,* Fred Stetson
- *Medicine for Mountaineering,* James A. Wilkerson, The Mountaineers (Seattle, WA)
- Mother Earth News, all back copies
- *The Natural Foods Epicure,* Nancy Albright
- *Shelters, Shacks, and Shanties,* D. C. Beard, Loompanics Unlimited
- *Survival (Field Manual 21–76),* U. S. Army manual
- *Survival at Sea,* R. P. Brandt, Paladin Press

APPENDIX

- *Survivalist's Medicine Chest,* Ragnar Benson, Paladin Press
- *Tools and How to Use Them,* Albert Jackson and David Day
- *Urban Alert,* Mary Ellen Clayton, Paladin Press
- *Wood Heat,* John Vivian
- *Work Horse Handbook,* Lynn R. Miller